The **BEST**
WRITING on
MATHEMATICS

2012

The **BEST** **WRITING** on **MATHEMATICS**

2012

Mircea Pitici, Editor

FOREWORD BY
DAVID MUMFORD

PRINCETON UNIVERSITY PRESS
PRINCETON AND OXFORD

For my parents

Contents

Foreword: The Synergy of Pure and Applied Mathematics,
of the Abstract and the Concrete
DAVID MUMFORD   ix

Introduction
MIRCEA PITICI   xvii

Why Math Works
MARIO LIVIO   1

Is Mathematics Discovered or Invented?
TIMOTHY GOWERS   8

The Unplanned Impact of Mathematics
PETER ROWLETT   21

An Adventure in the Nth Dimension
BRIAN HAYES   30

Structure and Randomness in the Prime Numbers
TERENCE TAO   43

The Strangest Numbers in String Theory
JOHN C. BAEZ AND JOHN HUERTA   50

Mathematics Meets Photography: The Viewable Sphere
DAVID SWART AND BRUCE TORRENCE   61

Dancing Mathematics and the Mathematics of Dance
SARAH-MARIE BELCASTRO AND KARL SCHAFFER   79

Can One Hear the Sound of a Theorem?
ROB SCHNEIDERMAN   93

Flat-Unfoldability and Woven Origami Tessellations
ROBERT J. LANG   113

A Continuous Path from High School Calculus to University Analysis
TIMOTHY GOWERS   129

Mathematics Teachers' Subtle, Complex Disciplinary Knowledge
    BRENT DAVIS   135

How to Be a Good Teacher Is an Undecidable Problem
    ERICA FLAPAN   141

How Your Philosophy of Mathematics Impacts Your Teaching
    BONNIE GOLD   149

Variables in Mathematics Education
    SUSANNA S. EPP   163

Bottom Line on Mathematics Education
    DAVID MUMFORD AND SOL GARFUNKEL   173

History of Mathematics and History of Science Reunited?
    JEREMY GRAY   176

Augustus De Morgan behind the Scenes
    CHARLOTTE SIMMONS   186

Routing Problems: A Historical Perspective
    GIUSEPPE BRUNO, ANDREA GENOVESE, AND GENNARO IMPROTA   197

The Cycloid and Jean Bernoulli
    GERALD L. ALEXANDERSON   209

Was Cantor Surprised?
    FERNANDO Q. GOUVÊA   216

Why Is There Philosophy of Mathematics at All?
    IAN HACKING   234

Ultimate Logic: To Infinity and Beyond
    RICHARD ELWES   255

Mating, Dating, and Mathematics: It's All in the Game
    MARK COLYVAN   262

Contributors   273

Notable Texts   281

Acknowledgments   285

Credits   287

Foreword: The Synergy of Pure and Applied Mathematics, of the Abstract and the Concrete

DAVID MUMFORD

All of us mathematicians have discovered a sad truth about our passion: It is pretty hard to tell anyone outside your field what you are so excited about! We all know the sinking feeling you get at a party when an attractive person of the opposite sex looks you in the eyes and asks—"What is it you do?" Oh, for a simple answer that moves the conversation along.

Now Mircea Pitici has stepped up to the plate and for the third year running has assembled a terrific collection of answers to this query. He ranges over many aspects of mathematics, including interesting pieces on the history of mathematics, the philosophy of mathematics, mathematics education, recreational mathematics, and even *actual presentations of mathematical ideas!* This volume, for example, has accessible discussions of n-dimensional balls, the intricacies of the distribution of prime numbers, and even of octonions (a strange type of algebra in which the "numbers" are 8-tuples of the ordinary sort of number)—none of which are easy to convey to the layperson. In addition—and I am equally pleased with this—several pieces explain in depth how mathematics can be used in science and in our lives—in dancing, for the traveling salesman, in search of marriage, and for full-surround photography, for instance.

To the average layperson, mathematics is a mass of abstruse formulae and bizarre technical terms (e.g., perverse sheaves, the monster group, barreled spaces, inaccessible cardinals), usually discussed by academics in white coats in front of a blackboard covered with peculiar symbols. The distinction between mathematics and physics is blurred and that between pure and applied mathematics is unknown.

But to the professional, these are three different worlds, different sets of colleagues, with different goals, different standards, and different customs.

The layperson has a point, though. Throughout history many practitioners have crossed seamlessly between one or another of these fields. Archimedes not only calculated the volume of a ball of radius r (a pure mathematics problem with the answer $4\pi r^3/3$) but also studied the lever (a physics problem) and used it both in warfare (applied mathematics: hurling fiery balls at Roman ships) and in mind experiments ("Give me a place to stand and I will move the earth"). Newton was both a brilliant mathematician (inventing calculus) and physicist (discovering the law of gravity).

Today it is different: The three fields no longer form a single space in which scientists can move easily back and forth. Starting in the mid-twentieth century, mathematicians were blindsided by the creation of quantum field theory and even more by string theory. Here physicists, combining their physical intuition with all the latest and fanciest mathematical theories, began to use mathematics in ways mathematicians could not understand. They abandoned rigorous reasoning in favor of physical intuition and played wildly with heuristics and extrapolations from well-known mathematics to "explain" the world of high energy. At about the same time (during the '50s and '60s), mathematics split into pure and applied camps. One group fell in love with the dream of a mathematics that lived in and for itself, in a Platonic world of blinding beauty. The English mathematician G. H. Hardy even boasted that his work could never be used for practical purposes. On the other side, another group wanted a mathematics that could solve real-world problems, such as defeating the Nazis. John von Neumann went to Los Alamos and devised a radical new type of mathematics based on gambling, the Monte Carlo technique, for designing the atom bomb. A few years later, this applied group developed a marvelous new tool, the computer—and with it applied mathematics was off and running in its own directions.

I have been deeply involved with both pure mathematics and applied mathematics. My first contact with real mathematical problems was during a summer job in 1953, when I used an analog computer to simulate the neutron flux in the core of an atomic reactor. I was learning the basics of calculus at the time, just getting used to writing Greek letters for numbers and operations—and the idea of connecting resistors in a grid to simulate Δ (technically, the Laplace differential operator) struck me as profoundly beautiful. I was struggling to get my mind around the abstract notions, but luckily I was well acquainted with the use of a soldering iron. I was delighted that I could construct simple electrical circuits that made calculus so tangible.

Later, in college, I found that I could not understand what quantum field theory was all about; *ergo*, I was not a physicist but a mathematician! I went all the way and immersed myself in one of the purest areas of pure mathematics. (One can get carried away: At one time the math department at Cambridge University advertised an opening and a misprint stated that the position was in the Department of "Purer" Mathematics!) I "constructed" something called "moduli schemes." I do not expect the reader to have ever heard of moduli schemes or have a clue what they are. But here is the remarkable thing: To mathematicians who study them, moduli schemes are just as real as the regular objects in the world.

I can explain at least the first steps of the mental gymnastics that led to moduli schemes. The key idea is that an ordinary object can be studied using *the set of functions on the object*. For example, if you have a pot of water, the water at each precise location, at each spatial point inside the pot, has a temperature. So temperature defines a function, a rule that associates to each point in the pot the real number that is the temperature at that exact point. Or you can measure the coordinates of each point, for instance, how many centimeters the point is above the stove. Secondly, you can do algebra with these functions—that is, you can add or multiply two such functions and get a third function. This step makes the set of these functions into a *ring*. I have no idea why, but when you have any set of things that can be added and multiplied, consistent with the usual rules (for instance, the distributive law a [times] $(b + c) = a$ [times] $b + a$ [times] c), mathematicians call this set a ring. You see, ordinary words are used in specialized ways. In our case, the ring contains all the information needed to describe the geometry of the pot because the points in the pot can be described by the map carrying each function to its value at that point.

Then the big leap comes: If you start with any ring—that is, any set of entities that can be added and multiplied subject to the usual rules, you simply and brashly declare that this creates a new kind of geometric object. The points of the object can be given by maps from the ring to the real numbers, as in the example of the pot. But they may also be given by maps to other *fields*. A field is a special sort of ring in which division is possible. To see how strange the situation becomes, the set consisting of just the numbers 0 and 1 with the rule $1 + 1 = 0$ is a field. As you see, pure mathematics revels in creating variations on the algebra and geometry you learned in high school. I have sometimes described the world that opens up to devotees as a secret garden for which you have to work hard before you get a key.

Applied mathematics is different. It is driven by real-world problems. You may want a mathematical model that accurately describes and predicts the fission of uranium in a nuclear reactor—my summer job in 1953—but there is no limit to the important practical problems to which mathematics can be applied, such as global warming, tornadoes, or tsunamis. Modeling these physical effects requires state-of-the-art mathematical tools known as partial differential equations (or PDEs)—an eighteenth century calculus invention. Or take biology and evolution; the folding of proteins, the process by which a neuron transmits information, and the evolution of new species have all led to major mathematical advances.

The reader now sees how easy it has been for pure and applied mathematics to drift apart. One group of practitioners are immersed in abstract worlds where all the rules have to be painstakingly guessed and proven, with no help from real life experience. The other is in constant touch with scientists and engineers and has to keep up with new data and new experiments. Their goal is always to make the right mathematical model, which captures the essential features of some practical situation, usually by simplifying the messiness of reality and often replacing rigorous derivations by numerical simulations. The example I mentioned above, of von Neumann's work on the atomic and hydrogen bombs, illustrates this approach. Von Neumann and his colleagues started by trying to model the explosion of the bomb using conventional PDEs. At a certain point, working with Nicholas Metropolis and Stan Ulam, his colleagues at Los Alamos, they had an inspiration: Let's imagine that a hundred neutrons in the bomb are gambling. Here gambling means that each of them collides with a uranium atom when the roll of the dice comes out right. You set up your dice so that its odds mimic those of the real neutron moving at the same speed. This game is a lot easier than following the approximately 1,000,000,000,000,000,000,000,000,000 neutrons that are really whizzing about—and it turned out to work well, unless you regret the legacy von Neumann's inspiration left the world.

But the drifting apart of pure and applied mathematics is not the whole story. The two worlds are tied more closely than you might imagine. Each contributes many ideas to the other, often in unexpected ways. Perhaps the most famous example is Einstein's need of new mathematical tools to push to deeper levels the ideas of special relativity. He found that Italian mathematicians, dealing with abstract n-dimensional space, had discovered tools for describing higher dimensional versions of curvature and the equations for shortest paths, called geodesics. Adapting these ideas, Einstein turned them into the foundations of general relativity (without which your global positioning system [GPS] wouldn't work). In the other direction, almost a century after Einstein discovered general relativity, working out the implications of Einstein's model is a hot area in pure mathematics, driving the invention of new techniques to deal with the highly nonlinear PDEs underlying his theory. In other words, pure mathematics made Einstein's physics possible, which in turn opened up new fields for pure mathematics.

A spectacular recent example of the interconnections between pure and applied mathematics involves prime numbers. No one (especially G. H. Hardy, as I mentioned) suspected that prime numbers could ever be useful in the real world, yet they are now the foundation of the encryption techniques that allow online financial transactions. This application is a small part of an industry of theoretical work on new algorithms for discrete problems—in particular, their classification by the order of magnitude of their speed—which is the bread and butter of computer science.

I want to describe another example of the intertwining of pure and applied mathematics in which I was personally involved. Computer vision research concerns writing computer code that will interpret camera and video input as effectively as humans can with their brains, by identifying all the objects and actions present. When this problem was first raised in the 1960s, many people believed that it was a straightforward engineering problem and would be solved in a few years. But fifty years later, computers still cannot recognize specific individuals from their faces very well or name the objects and read all the signs in a street scene. We are getting closer: Computers are pretty good at least at finding all the faces in a scene, if not identifying who they are. A computer can even drive a car (at least when the traffic isn't too bad).

Several things have been crucial for this progress. The first thing was the recognition that visual analysis is not a problem of deduction—i.e., combining the rules of logic with a set of learned rules about the nature of the objects that fill our world. It turns out that our knowledge is always too incomplete and our visual data is too noisy and cluttered to be interpreted by deduction. In this situation, the method of reasoning needed to parse a real-world scene must be statistical, not deductive. To implement this form of reasoning, our knowledge of the world must be encoded in a probabilistic form, known as an *a priori* probability distribution. This distribution tells you things like this: The likelihood of seeing a tiger walk around the corner is smaller than that of seeing a dog walk around the corner. It is called *a priori* because we know this data before we start to analyze the scene present now to our eyes. Analyzing noisy, incomplete data using *a priori* knowledge is called Bayesian inference, after the Reverend Thomas Bayes, who proposed this form of statistical analysis in the eighteenth century.

But what kinds of probability distributions are going to be used? Here, computer vision drew on a wide variety of mathematical tools and, in turn, stimulated the development of new variants and new algorithms in many fields of mathematics and physics. In particular, statistical mechanics contributed a tool known as Gibbs probability models and a variety of techniques for analyzing them. The conversion of an image into a cartoon, in which the main objects are outlined, turns out to have much in common with a set of pure math problems called "free boundary value problems." These are problems that call for solving for an unknown and changing boundary between two distinct areas or volumes, such as a melting ice cube in water. For instance, analyzing an MRI to see if an organ of the body is diseased or normal has stimulated work in the mathematics of infinite dimensional spaces. This is because from a mathematical viewpoint, the set of possible shapes of the organ is best studied as the set of points in a space; there are infinitely many ways in which shape can vary, so this space must have infinitely many dimensions.

I hope I have convinced you that one of the striking features of the spectrum of related fields—pure mathematics, applied mathematics, and physics—is how unexpected connections are always being discovered. I talked about a variety of such connections *among* these fields. But even within pure mathematics, amazing connections between remote areas are uncovered all the time. In the last decade, for example, ideas from number theory have led to progress in the understanding of the topology of high-dimensional spheres.

It will be difficult to fully repair the professional split between pure and applied mathematics and between mathematics and physics. One reason for this difficulty is that each academic field has grown so much, so that professionals have limited time to read work outside their specialties. It is not easy to master more than a fraction of the work in any single field, let alone in more than one. What we need, therefore, is to work harder at explaining our work to each other. This book, though it is addressed mostly to lay people, is a step in the right direction.

As I see it, the major obstacle is that there are two strongly conflicting traditions of writing and lecturing about mathematics. In pure mathematics (but not exclusively), the twentieth century saw the development of an ideal exposition as one that started at the most abstract level and then gradually narrowed the focus. This style was especially promoted by the French writing collaborative "Bourbaki." In the long tradition of French encyclopedists, the mathematicians forming the Bourbaki group sought to present the entire abstract structure of all mathematical concepts in one set of volumes, the *Éléments de Mathématique*. In that treatise the real numbers, which most of us regard as a starting point, only appeared midway into the series as a special "locally compact topological field." In somewhat less relentless forms, their orientation has affected a large proportion of all writing and lecturing in mathematics.

An opposing idea, promoted especially in the Russian school, is that a few well-chosen examples can illuminate an entire field. For example, one can learn stochastic processes by starting with a simple random walk, moving on to Brownian motion, its continuous version, and then to more abstract and general processes. I remember a wonderful talk on hyperbolic geometry by the mathematician Bill Thurston, where he began by scrawling with yellow chalk on the board: He explained that it was a simple drawing of a fire. His point was that in hyperbolic space, you have to get much, much closer to the fire to warm up than you do in Euclidean space. Along with such homey illustrations, there is also the precept "lie a little." If we insist on detailing all the technical qualifications of a theorem, we lose our readers or our audience very fast. If we learn to say things simply and build up slowly from the concrete to the abstract, we may be able to build many bridges among our various specialties. For me, this style will always be *The Best Writing on Mathematics,* and this book is full of excellent examples of it.

Introduction

MIRCEA PITICI

A little more than eight years ago I planned a series of "best writing" on mathematics with the sense that a sizable and important literature does not receive the notice, the consideration, and the exposure it deserves. Several years of thinking on such a project (for a while I did not find a publisher interested in my proposal) only strengthened my belief that the best of the nontechnical writings on mathematics have the potential to enhance the public reception of mathematics and to enrich the interdisciplinary and intradisciplinary dialogues so vital to the emergence of new ideas.

The prevailing view holds that the human activity we conventionally call "mathematics" is mostly beyond fruitful debate or personal interpretation because of the uncontested (and presumably uncontestable) matters of fact pertaining to its nature. According to this view, mathematics speaks for itself, through its cryptic symbols and the efficacy of its applications.

At close inspection, the picture is more complicated. Mathematics has been the subject of numerous disputes, controversies, and crises—and has weathered them remarkably well, growing from the resolution of the conundrums that tested its strength. By doing so, mathematics has become a highly complex intellectual endeavor, thriving at the ever-shifting intersection of multiple polarities that can be used to describe its characteristics. Consequently, most people who are engaged with mathematics (and many people disengaged from it) do it on a more personal level than they are ready to admit. Writing is an effective way of informing others on such individualized positioning vis-à-vis mathematics. A growing number of authors—professionals and amateurs—are taking on such a task. Every week new books on mathematics are published, in a dazzling blossoming of the genre hard to imagine even a decade ago (I mention a great number of these titles later in this introduction). This recent flourishing confirms that, just as mathematics offers unlimited possibilities for asking new questions, formulating new problems, opening new theoretical vistas, and rethinking old concepts, narrating our individualized perspectives on it is equally potent in expressivity and in impact.

By editing this annual series, I stand for the wide dissemination of insightful writings that touch on any aspect related to mathematics. I aim to diminish the gap between mathematics professionals and the general public and to give exposure to a substantial literature that is not currently used systematically in scholarly settings. Along the way, I hope to weaken or even to undermine some of the barriers that stand between mathematics and its pedagogy, history, and philosophy, thus alleviating the strains of hyperspecialization and offering opportunities for connection and collaboration among people involved with different aspects of mathematics. If, by presenting in each volume a snapshot of contemporary thinking on mathematics, we succeed in building a useful historical reference, in offering an informed source of further inquiry, and in encouraging even more exceptional writing on mathematics, so much the better.

Overview of the Volume

In the first article of our selection, Mario Livio ponders the old question of what makes mathematics effective in describing many features of the physical universe and proposes that its power lies in the peculiar blending between the human ingenuity in inventing flexible and adaptable mathematical tools and the uncanny regularities of the universe.

Timothy Gowers brings the perspective of a leading research mathematician to another old question, that concerning features of discovery and elements of invention in mathematics; he discusses some of the psychological aspects of this debate and illustrates it with a wealth of examples.

In a succession of short pieces, Peter Rowlett and his colleagues at the British Society for the History of Mathematics present the unexpected applications, ricocheting over centuries, of notions and results long believed to have no use beyond theoretical mathematics.

Brian Hayes puzzles over the proportion between the volume of a sphere and that of the cube circumscribed to it, in various dimensions—and offers cogent explanations for the surprising findings that this proportion reaches a maximum in five dimensions and it decreases rapidly to insignificant values as the number of dimensions increases.

Terence Tao tackles a few conundrums concerning the distribution of prime numbers, noting that the distribution displays elements of order *and* of chaos, thus challenging simplistic attempts to elucidate its patterns.

John C. Baez and John Huerta tell the story of the tenacious William R. Hamilton in search of a better number system—and how his invention of the quaternions led John Graves and Arthur Cayley (independently) to thinking up the octonions, which play a vital role in the theory of strings.

David Swart and Bruce Torrence discuss several ways of projecting a sphere on a plane, the qualitative trade-offs involved in them, the basic mathematics underlying such correspondence, and their applications to panoramic photography.

Drawing on their experience as mathematicians and choreographers, sarah-marie belcastro and Karl Schaffer explore the interplay of dancing and such mathematical ideas as symmetry, group structure, topological links, graphs, and (suggestions of) infinity.

A musician and mathematician, Rob Schneiderman, offers a critical viewpoint on the literature that explores the hidden and the overt connections between music and mathematics, pleading for a deepening of the discourse on their interaction.

Robert J. Lang describes a mathematical condition ensuring that an origami paper folding unfolds into a flat piece of paper.

In the second of two texts in this volume, Timothy Gowers argues that high-school level mathematical thinking of calculus notions is not as different as most professionals assume from the thought process common to similar notions of analysis taught in university courses, at a higher level of rigor and abstraction.

Brent Davis argues that successful teaching of mathematics relies more on tacit, unconscious skills that support the learner's full engagement with mathematics, than on the formal training currently available in preparatory programs for teachers.

Erica Flapan tells us, with disarming candor, about her charming adventures in search of the best methods for teaching mathematics—and concludes that, despite various degrees of success, she remains an agnostic in this matter.

Bonnie Gold argues that everyone teaching mathematics does it according to certain philosophical assumptions about the nature of mathematics—whether the assumptions are explicit or remain implicit.

Susanna S. Epp examines several uses of the concept of "variable" in mathematics and opines that, from an educational standpoint, the best is to treat variables as placeholders for numerals.

David Mumford and Sol Garfunkel plead for a broad reform of the U.S. system of mathematics education, more attuned to the practical uses of mathematics for the citizenry and less concerned with the high-stakes focus on testing currently undertaken in the United States.

Jeremy Gray surveys recent trends in the study of the history of mathematics as compared to research on the history of science and examines the possibility that the two might be somehow integrated in the future.

Charlotte Simmons writes about Augustus De Morgan as a mentor of other mathematicians, an aspect less known than the research contributions of the great logician.

Giuseppe Bruno, Andrea Genovese, and Gennaro Improta review several formulations of various routing problems, with wide applications to matters of mathematical optimization.

Special curves were at the forefront of mathematical research about three centuries ago, and one of them, the cycloid, attracted the attention (and the rivalry) of the most famous mathematicians of the time—as Gerald L. Alexanderson shows in his piece on the Bernoulli family.

Fernando Gouvêa examines Georg Cantor's correspondence, to trace the original meaning of Cantor's famous remark "I see it, but I don't believe it!" and to refute the ulterior, psychological interpretations that other people have given to this quip.

Ian Hacking explains that the enduring fascination and the powerful influence of mathematics on so many Western philosophers lie in the experiences engendered on them by learning and doing mathematics.

Richard Elwes delves into the subtleties of mathematical infinity and ventures some speculations on the future clarification of the problems it poses.

Finally, Mark Colyvan illustrates the basic mathematics involved in the games of choice we encounter in life, whenever we face processes that require successive alternative decisions.

Other Notable Writings

As in previous years, I selected the texts in this volume from a much larger group of articles. At the end of the book is a list of other remarkable pieces that I considered but did not include for reasons of space or related to copyright. In this section of the introduction, I mention a string of books on mathematics that came to my attention over the past year.

It is fit to start the survey of recent nontechnical books on mathematics by mentioning a remarkable reference work that fills a gap in the literature about mathematics, the *Encyclopedia of Mathematics and Society*, edited in three massive volumes by Sarah J. Greenwald and Jill E. Thomley.

A few new books inform pertinently on mathematicians' lives, careers, and experiences. The *Academic Genealogy of Mathematicians* by Soo-young Chang is impressive; Donald J. Albers and Gerald L. Alexanderson follow up on previous volumes of interviews with *Fascinating Mathematical People*; and Amir D. Aczel presents a highly readable collection of biographies of famous mathematicians in *A Strange Wilderness*. Other historical biographies touch on more than immediate life details, by examining the broader influence of their subjects: *Remembering Sofya Kovalevskaya* by Michèle Audin, *Abraham De Moivre* by David Bellhouse, *Turbulent Times in Mathematics* by Elaine McKinnon Riehm and Frances Hoffman (on J. C. Fields), *The Man of Numbers* by Keith Devlin (on Fibonacci), *Giuseppe Peano between Mathematics and Logic* edited by Fulvia Skof, and *Stefan Banach*, a collection edited by Emilia Jakimowicz and Adam Miranowicz.

Several highly accessible books of popular mathematics introduce the reader to a potpourri of basic and advanced mathematical notions or to encounters with mathematics in daily life. They include Peter M. Higgins's *Numbers*, James D. Stein's *Cosmic Numbers*, Marcus Du Sautoy's *The Number Mysteries*, Ian Stewart's *The Mathematics of Life*, David Berlinski's *One, Two, Three*, and Tony Crilly's *Mathematics*. Slightly more technical and focused on particular topics are William J. Cook's *In Pursuit of the Traveling Salesman*, Alfred S. Posamentier and Ingmar Lehmann's *The Glorious Golden Ratio*, and *Probability Tales* by Charles M. Grinstead and collaborators. An elegant book in an unusual format is Nicolas Bouleau's *Risk and Meaning*. And an accessible path to higher geometry is in *Geometry Revealed* by Marcel Berger.

The number of interdisciplinary and applicative books that build connections between mathematics and other domains continues to grow fast. On mathematics and music, we recently have *A Geometry of Music* by Dmitri Tymoczko and *The Science of String Instruments,* edited by Thomas D. Rossing. Some remarkable books on mathematics and architecture are now available, including *The Function of Form* by Farshid Moussavi (marvelously illustrated); *Advances in Architectural Geometry 2010,* edited by Cristiano Ceccato and his collaborators; *The New Mathematics of Architecture* by Jane and Mark Burry; the 30th anniversary reissue of *The Dynamics of Architectural Form* by Rudolf Arnheim; and *Matter in the Floating World,* a book of interviews by Blaine Brownell.

Among the books on mathematics and other sciences are Martin B. Reed's *Core Maths for the Biosciences*; *BioMath in the Schools,* edited by Margaret B. Cozzens and Fred S. Roberts; *Chaos: The Science of Predictable Random Motion* by Richard Kautz (a historical overview); *Some Mathematical Models from Population Genetics* by Alison Etheridge; and *Mathematics Meets Physics,* a collection of historical pieces (in English and German) edited by Karl-Heinz Schlote and Martina Schneider.

Everyone expects some books on mathematics and social sciences; indeed, this time we have *Mathematics of Social Choice* by Christoph Börgers, *Bond Math* by Donald J. Smith, *An Elementary Introduction to Mathematical Finance* by Sheldon M. Ross, and *E. E. Slutsky as Economist and Mathematician* by Vincent Barnett. A highly original view on mathematics, philosophy, and financial markets is *The Blank Swan* by Elie Ayache. And an important collection of papers concerning statistical judgment in the real world is David A. Freedman's *Statistical Models and Causal Inference.*

More surprising reaches of mathematics can be found in *Magical Mathematics* by Persi Diaconis and Ron Graham, *Math for the Professional Kitchen* (with many worksheets for your convenience) by Laura Dreesen, Michael Nothnagel, and Susan Wysocki, *The Hidden Mathematics of Sport* by Rob Eastaway and John Haigh, *Face Geometry and Appearance Modeling* by Zicheng Liu and Zhengyou Zhang, *How to Fold It* by Joseph O'Rourke, and *Mathematics for the Environment* by Martin E. Walter. Sudoku comes of (mathematical) age in *Taking Sudoku Seriously* by Jason Rosenhouse and Laura Taalman. More technical books, but still interdisciplinary and accessible, are *Viewpoints: Mathematical Perspective and Fractal Geometry in Art* by Marc Frantz and Annalisa Crannell and *Infinity: New Research Frontiers,* edited by Michael Heller and W. Hugh Woodin.

Many new books have been published recently in mathematics education, too many to mention them all. Several titles that caught my attention are Tony Brown's *Mathematics Education and Subjectivity*, Hung-Hsi Wu's unlikely voluminous *Understanding Numbers in Elementary School Mathematics*, Judith E. Jacobs' *A Winning Formula for Mathematics Instruction*, as well as *Upper Elementary Math Lessons* by Anna O. Graeber and her collaborators, *The Shape of Algebra in the Mirrors of Mathematics* by Gabriel Katz and Vladimir Nodelman, and *Geometry: A Guide for Teachers* by Judith and Paul Sally. Keith Devlin offers an original view of the connections between computer games and mathematics learning in *Mathematics Education for a New Era*. Among the many volumes at the National Council of Teachers of Mathematics, notable is the 73rd NCTM Yearbook, *Motivation and Disposition*, edited by Daniel J. Brahier and William R. Speer; *Motivation Matters and Interest Counts* by James Middleton and Amanda Jansen; and *Disrupting Tradition* by William Tate and colleagues. NCTM also publishes many books to support the professional development of mathematics teachers. A good volume for preschool teachers is *Math from Three to Seven* by Alexander Zvonkin. With an international perspective are *Russian Mathematics Education,* edited by Alexander Karp and Bruce R. Vogeli; *International Perspectives on Gender and Mathematics Education,* edited by Helen J. Forgasz and her colleagues; and *Teacher Education Matters* by William H. Schmidt and his colleagues. *Mathematics Teaching and Learning Strategies in PISA*, published by the Organisation for Economic Co-operation and Development, contains a wealth of statistics on global mathematics education.

An excellent volume at the intersection of brain research, psychology, and education, with several contributions focused on learning mathematics, is *The Adolescent Brain,* edited by Valerie F. Reyna and her collaborators.

Besides the historical biographies mentioned above, several other contributions to the history of mathematics are worth enumerating. Among thematic histories are Ranjan Roy's *Sources in the Development of Mathematics*, a massive and exhaustive account of the growth of the theory of series and products; *Early Days in Complex Dynamics* by Daniel S. Alexander and collaborators; *The Origin of the Logic of Symbolic Mathematics* by Burt C. Hopkins; *Lobachevski Illuminated* by Seth Braver; *Mathematics in Victorian Britain,* edited by Raymond Flood and collaborators; *Journey through Mathematics* by Enrique A. González-Velasco; and

Histories of Computing by Michael Sean Mahoney. Two remarkable books that weave the history of mathematics and European arts are *Between Raphael and Galileo* by Alexander Marr and *The Passionate Triangle* by Rebecca Zorach.

Several historical editions are newly available, for instance, Lobachevsky's *Pangeometry*, translated and edited by Athanase Papadopoulos; *80 Years of Zentralblatt MATH*, edited by Olaf Teschke and collaborators; and Albert Lautman's *Mathematics, Ideas, and the Physical Real*. Other historical works are *The Theory That Would Not Die* by Sharon Bertsch McGrayne, *From Cardano's Great Art to Lagrange's Reflections* by Jacqueline Stedall, *Chasing Shadows* by Clemency Montelle, and *World in the Balance* by Robert P. Crease.

In philosophy of mathematics, a few books concern personalities: *After Gödel* by Richard Tieszen; *Kurt Gödel and the Foundations of Mathematics*, edited by Matthias Baaz et al.; *Spinoza's Geometry of Power* by Valtteri Viljanen; *Bolzano's Theoretical Philosophy* by Sandra Lapointe; and *New Essays on Peirce's Mathematical Philosophy*, edited by Matthew E. Moore. Other recent volumes on the philosophy of mathematics and its history are Paolo Mancosu's *The Adventure of Reason*, Paul M. Livingston's *The Politics of Logic*, Gordon Belot's *Geometric Possibility*, and *Fundamental Uncertainty*, edited by Silva Marzetti Dall'Aste Brandolini and Roberto Scazzieri.

Mathematics meets literature in William Goldbloom Bloch's *The Unimaginable Mathematics of Borges' Library of Babel* and, in a different way, in *All Cry Chaos* by Leonard Rosen (where the murder of a mathematician is pursued by a detective called Henri Poincaré). Hans Magnus Enzensberger, the German writer who authored the very successful book *The Number Devil*, has recently published the tiny booklet *Fatal Numbers*.

For other titles the reader is invited to check the introduction to the previous volumes of *The Best Writing on Mathematics*.

As usual, at the end of the introduction I mention several interesting websites. A remarkable bibliographic source is the online list of references on Benford's Law organized by Arno Berger, Theodore Hill, and Erika Rogers (http://www.benfordonline.net/). Other good topic-oriented websites are the MacTutor History of Mathematics archive from the University of St. Andrews in Scotland (http://www-history.mcs .st-and.ac.uk/), Mathematicians of the African Diaspora (MAD) (http:// www.math.buffalo.edu/mad/), the Famous Curves index (http://www

-history.mcs.st-and.ac.uk/Curves/Curves.html), the National Curve Bank (http://curvebank.calstatela.edu/index/index.htm), and Free Mathematics Books (http://www.e-booksdirectory.com/mathematics.php). An intriguing site dedicated to the work of Alexandre Grothendieck, one of the most intriguing mathematicians alive, is the Grothendieck Circle (http://www.grothendieckcircle.org/). An excellent website for mathematical applications in science and engineering is Equalis (http://www.equalis.com/). Among websites with potential for finding materials for mathematical activities are the one on origami belonging to Robert Lang, a contributor to this volume (http://www.lang origami.com/index.php4); many other Internet sources for the light side of mathematics can be found conveniently on the personal page maintained by Greg Frederickson of Purdue University (http://www.cs .purdue.edu/homes/gnf/hotlist.html).

I hope you, the reader, find the same value and excitement in reading the texts in this volume as I found while searching, reading, and selecting them. For comments on this book and to suggest materials for consideration in preparing future volumes, I encourage you to send correspondence to me: Mircea Pitici, P.O. Box 4671, Ithaca, NY 14852.

Works Mentioned

Aczel, Amir D. *A Strange Wilderness: The Lives of the Great Mathematicians*. New York: Sterling, 2011.

Albers, Donald J., and Gerald L. Alexanderson. (Eds.) *Fascinating Mathematical People: Interviews and Memoirs*. Princeton, NJ: Princeton Univ. Press, 2011.

Alexander, Daniel S., Felice Iavernaro, and Alessandro Rosa. *Early Days in Complex Dynamics: A History of Complex Dynamics in One Variable During 1906–1942*. Providence, RI: American Mathematical Society, 2011.

Arnheim, Rudolf. *The Dynamics of Architectural Form*. 30th anniversary ed., Berkeley, CA: Univ. of California Press, 2009.

Audin, Michèle. *Remembering Sofya Kovalevskaya*. Heidelberg, Germany: Springer Verlag, 2011.

Ayache, Elie. *The Blank Swan: The End of Probability*. Chichester, UK: Wiley, 2010.

Baaz, Matthias, et al. (Eds.) *Kurt Gödel and the Foundations of Mathematics: Horizons of Truth*. New York: Cambridge Univ. Press, 2011.

Barnett, Vincent. *E. E. Slutsky as Economist and Mathematician*. New York: Routledge, 2011.

Bellhouse, David R. *Abraham De Moivre: Setting the Stage for Classical Probability and Its Applications*. Boca Raton, FL: Taylor & Francis, 2011.

Belot, Gordon. *Geometric Possibility*. Oxford, UK: Oxford Univ. Press, 2011.

Berger, Marcel. *Geometry Revealed: A Jacob's Ladder to Modern Higher Geometry*. Heidelberg, Germany: Springer Verlag, 2011.

Berlinski, David. *One, Two, Three: Absolutely Elementary Mathematics.* New York: Pantheon Books, 2011.

Bloch, William Goldbloom. *The Unimaginable Mathematics of Borges' Library of Babel.* Oxford, UK: Oxford Univ. Press, 2008.

Börgers, Christoph. *Mathematics of Social Choice: Voting, Compensation, and Division.* Philadelphia, PA: SIAM, 2011.

Bouleau, Nicolas. *Risk and Meaning: Adversaries in Art, Science and Philosophy.* Berlin, Germany: Springer Verlag, 2011.

Brahier, Daniel J., and William R. Speer. (Eds.) *Motivation and Disposition: Pathways to Learning Mathematics.* Reston, VA: The National Council of Teachers of Mathematics, 2011.

Brandolini, Silva Marzetti Dall'Aste, and Roberto Scazzieri. (Eds.) *Fundamental Uncertainty: Rationality and Plausible Reasoning.* New York: Palgrave MacMillan, 2011.

Braver, Seth. *Lobachevski Illuminated.* Washington, DC: Mathematical Association of America, 2011.

Brown, Tony. *Mathematics Education and Subjectivity: Cultures and Cultural Renewal.* Heidelberg, Germany: Springer Verlag, 2011.

Brownell, Blaine. *Matter in the Floating World: Conservations with Leading Japanese Architects and Designers.* New York: Princeton Architectural Press, 2011.

Burry, Jane, and Mark Burry. *The New Mathematics of Architecture.* New York: Thames & Hudson, 2010.

Ceccato, Cristiano, et al. (Eds.) *Advances in Architectural Geometry 2010.* Vienna, Austria: Springer Verlag, 2010.

Chang, Sooyoung. *Academic Genealogy of Mathematicians.* Singapore: World Scientific, 2011.

Cook, William J. *In Pursuit of the Traveling Salesman: Mathematics at the Limits of Computation.* Princeton, NJ: Princeton Univ. Press, 2012.

Cozzens, Margaret B., and Fred S. Roberts. (Eds.) *BioMath in the Schools.* Providence, RI: American Mathematical Society, 2011.

Crease, Robert P. *World in the Balance: The Historic Quest for an Absolute System of Measurement.* New York: W. W. Norton, 2011.

Crilly, Tony. *Mathematics. The Big Questions.* New York: Metro Books, 2011.

Devlin, Keith. *The Man of Numbers: Fibonacci's Arithmetic Revolution.* New York: Walker Publishing, 2011.

Devlin, Keith. *Mathematics Education for a New Era: Video Games as a Medium for Learning.* Natick, MA: A. K. Peters, 2011.

Diaconis, Persi, and Ron Graham. *Magical Mathematics: The Mathematical Ideas That Animate Great Magic Tricks.* Princeton, NJ: Princeton Univ. Press, 2011.

Dreesen, Laura, Michael Nothnagel, and Susan Wysocki. *Math for the Professional Kitchen.* Hoboken, NJ: John Wiley & Sons, 2011.

Du Sautoy, Marcus. *The Number Mysteries: A Mathematical Odyssey through Everyday Life.* New York: Palgrave MacMillan, 2011.

Eastaway, Rob, and John Haigh. *The Hidden Mathematics of Sport.* London: Portico, 2011.

Enzensberger, Hans Magnus. *Fatal Numbers: Why Count on Chance.* Translated by Karen Leeder. New York: Upper West Side Philosophers, 2011.

Etheridge, Alison. *Some Mathematical Models from Population Genetics.* London: Springer Verlag, 2011.

Flood, Raymond, Adrian Rice, and Robin Wilson. (Eds.) *Mathematics in Victorian Britain.* Oxford, UK: Oxford Univ. Press, 2011.

Forgasz, Helen J., Joanne Rossi Becker, Kyeong-Hwa Lee, and Olof Bjorg Steinthorsdottir. (Eds.) *International Perspectives on Gender and Mathematics Education.* Charlotte, NC: Information Age Publishing, 2010.

Frantz, Marc, and Annalisa Crannell. *Viewpoints: Mathematical Perspectives on Fractal Geometry in Art*. Princeton, NJ: Princeton Univ. Press, 2011.

Freedman, David A. *Statistical Models and Causal Inference: A Dialogue with the Social Sciences*. New York: Cambridge Univ. Press, 2010.

González-Velasco, Enrique A. *Journey through Mathematics: Creative Episodes in Its History*. Heidelberg, Germany: Springer Verlag, 2011.

Graeber, Anna O., Linda Valli, and Kristie Jones Newton. *Upper Elementary Math Lessons*. Lanham, MD: Rowman & Littlefield, 2011.

Greenwald, Sarah J., and Jill E. Thomley. (Eds.) *Encyclopedia of Mathematics and Society*, 3 volumes. Pasadena, CA: Salem Press, 2011.

Grinstead, Charles M., William P. Peterson, and J. Laurie Snell. *Probability Tales*. Providence, RI: American Mathematical Society, 2011.

Heller, Michael, and W. Hugh Woodin. (Eds.) *Infinity: New Research Frontiers*. New York: Cambridge Univ. Press, 2011.

Higgins, Peter M. *Numbers: A Very Short Introduction*. Oxford, UK: Oxford Univ. Press, 2011.

Hopkins, Burt C. *The Origin of the Logic of Symbolic Mathematics: Edmund Husserl and Jacob Klein*. Bloomington, IN: Indiana Univ. Press, 2011.

Jacobs, Judith E. *A Winning Formula for Mathematics Instruction: Converting Research into Results*. Alexandria, VA: Educational Research Service, 2011.

Jakimowicz, Emilia, and Adam Miranowicz. (Eds.) *Stefan Banach: Remarkable Life, Brilliant Mathematics*. Gdańsk, Poland: Gdańsk Univ. Press, 2011.

Karp, Alexander, and Bruce R. Vogeli. (Eds.) *Russian Mathematics Education: Programs and Practices*. Singapore: World Scientific, 2011.

Katz, Gabriel, and Vladimir Nodelman. *The Shape of Algebra in the Mirrors of Mathematics*. Singapore: World Scientific, 2011.

Kautz, Richard. *Chaos: The Science of Predictable Random Motion*. Oxford, UK: Oxford Univ. Press, 2011.

Lapointe, Sandra. *Bolzano's Theoretical Philosophy: An Introduction*. New York: Palgrave, 2011.

Lautman, Albert. *Mathematics, Ideas, and the Physical Real*. London: Continuum, 2011.

Liu, Zicheng, and Zhengyou Zhang. *Face Geometry and Appearance Modeling: Concepts and Applications*. Cambridge, MA: Harvard Univ. Press, 2011.

Livingston, Paul M. *The Politics of Logic: Badiou, Wittgenstein, and the Consequences of Formalism*. New York: Routledge, 2012.

Lobachevsky, Nikolai I. *Pangeometry*. Edited and translated by Athanase Papadoupulos. Zürich, Switzerland: European Mathematical Society, 2010.

Mahoney, Michael Sean. *Histories of Computing*. Cambridge, MA: Harvard Univ. Press, 2011.

Mancosu, Paolo. *The Adventure of Reason: The Interplay between Philosophy of Mathematics and Mathematical Logic, 1900–1940*. Oxford, UK: Oxford Univ. Press, 2011.

Marr, Alexander. *Between Raphael and Galileo: Mutio Oddi and the Mathematical Culture of Late Renaissance Italy*. Chicago: Chicago Univ. Press, 2011

McGrayne, Sharon Bertsch. *The Theory That Would Not Die: How Bayes' Rule Cracked the Enigma Code, Hunted Down Russian Submarines, and Emerged Triumphant from Two Centuries of Controversies*. New Haven, CT: Yale Univ. Press, 2011.

Middleton, James A., and Amanda Jansen. *Motivation Matters and Interest Counts: Fostering Engagement in Mathematics*. Reston, VA: The National Council of Teachers of Mathematics, 2011.

Montelle, Clemency. *Chasing Shadows: Mathematics, Astronomy, and the Early History of Eclipse Reckoning*. Baltimore, MD: Johns Hopkins Univ. Press, 2011.

Moore, Matthew E. (Ed.) *New Essays on Peirce's Mathematical Philosophy*. La Salle, IL: Chicago Univ. Press, 2010.

Moussavi, Farshid. *The Function of Form*. Boston: Harvard Graduate School of Design, 2010.

Organisation for Economic Co-operation and Development. *Mathematics Teaching and Learning Strategies in PISA*. OECD, Paris: 2010.

O'Rourke, Joseph. *How to Fold It: The Mathematics of Linkages, Origami, and Polyhedra*. New York: Cambridge Univ. Press, 2011.

Posamentier, Alfred S., and Ingmar Lehmann. *The Glorious Golden Ratio*. Amherst, NY: Prometheus Books, 2012.

Reed, Martin B. *Core Maths for the Biosciences*. Oxford, UK: Oxford Univ. Press, 2011.

Reyna, Valerie, F., Sandra B. Chapman, Michael R. Dougherty, and Jere Confrey. (Eds.) *The Adolescent Brain: Learning, Reasoning, and Decision Making*. Washington, DC: American Psychological association, 2012.

Riehm, Elaine McKinnon, and Frances Hoffman. *Turbulent Times in Mathematics: The Life of J. C. Fields and the History of the Fields Medal*. Providence, RI: American Mathematical Society, 2011.

Rosen, Leonard. *All Cry Chaos*. Sag Harbor, NY: Permanent Press, 2011.

Rosenhouse, Jason, and Laura Taalman. *Taking Sudoku Seriously: The Math behind the World's Most Popular Pencil Puzzle*. Oxford, UK: Oxford Univ. Press, 2011.

Ross, Sheldon M. *An Elementary Introduction to Mathematical Finance*. New York: Cambridge Univ. Press, 2011.

Rossing, Thomas D. (Ed.) *The Science of String Instruments*. Heidelberg, Germany: Springer Verlag, 2011.

Roy, Ranjan. *Sources in the Development of Mathematics: Series and Products from the Fifteenth to the Twenty-First Century*. New York, NY: Cambridge Univ. Press, 2011.

Sally, Judith D., and Paul J. Sally, Jr. *Geometry: A Guide for Teachers*. Providence, RI: American Mathematical Society, 2011.

Schlote, Karl-Heinz, and Martina Schneider. (Eds.) *Mathematics Meets Physics: A Contribution to Their Interaction in the 19th and the First Half of the 20th Century*. Leipzig, Germany: Harri Deutsch, 2011.

Schmidt, William H., Sigrid Blömeke, and Maria Teresa Tatto. *Teacher Education Matters: A Study of Middle School Mathematics Teacher Preparation in Six Countries*. New York: Teachers College, Columbia Univ. Press, 2011.

Skof, Fulvia. (Ed.) *Giuseppe Peano between Mathematics and Logic*. London: Springer Verlag, 2011.

Smith, Donald J. *Bond Math: The Theory behind the Formulas*. New York: John Wiley & Sons, 2011.

Stedall, Jacqueline. *From Cardano's Great Art to Lagrange's Reflections: Filling a Gap in the History of Algebra*. Zürich, Switzerland: European Mathematical Society, 2011.

Stein, James D. *Cosmic Numbers: The Numbers That Define Our Universe*. New York, NY: Basic Books, 2011.

Stewart, Ian. *The Mathematics of Life*. New York: Basic Books, 2011.

Tate, William F., Karen D. King, and Celia Rousseau Anderson. *Disrupting Tradition: Research and Practice Pathways in Mathematics Education*. Reston, VA: The National Council of Teachers of Mathematics, 2011.

Teschke, Olaf, Bernd Wegner, and Dirk Werner. (Eds.) *80 Years of Zentralblatt MATH: 80 Footprints of Distinguished Mathematicians in Zentralblatt*. Heidelberg, Germany: Springer Verlag, 2011.

Tieszen, Richard. *After Gödel: Platonism and Rationalism in Mathematics and Logic*. Oxford, UK: Oxford Univ. Press, 2011.

Tymoczko, Dmitri. *A Geometry of Music: Harmony and Counterpoint in the Extended Common Practice*. Oxford, UK: Oxford Univ. Press, 2011.

Viljanen, Valtteri. *Spinoza's Geometry of Power.* Cambridge, UK: Cambridge Univ. Press, 2011.

Walter, Martin E. *Mathematics for the Environment.* Boca Raton, FL: Taylor & Francis, 2011.

Wu, Hung-Hsi. *Understanding Numbers in Elementary School Mathematics.* Providence, RI: American Mathematical Society, 2011.

Zorach, Rebecca. *The Passionate Triangle.* Chicago: Chicago Univ. Press, 2011.

Zvonkin, Alexander. *Math from Three to Seven: The Story of a Mathematical Circle for Preschoolers.* Providence, RI: American Mathematical Society, 2011.

The **BEST**
WRITING on
MATHEMATICS

2012

Why Math Works

Mario Livio

Most of us take it for granted that math works—that scientists can devise formulas to describe subatomic events or that engineers can calculate paths for spacecraft. We accept the view, espoused by Galileo, that mathematics is the language of science and expect that its grammar explains experimental results and even predicts novel phenomena. The power of mathematics, though, is nothing short of astonishing. Consider, for example, Scottish physicist James Clerk Maxwell's famed equations: not only do these four expressions summarize all that was known of electromagnetism in the 1860s, they also anticipated the existence of radio waves two decades before German physicist Heinrich Hertz detected them. Very few languages are as effective, able to articulate volumes worth of material so succinctly and with such precision. Albert Einstein pondered, "How is it possible that mathematics, a product of human thought that is independent of experience, fits so excellently the objects of physical reality?"

As a working theoretical astrophysicist, I encounter the seemingly "unreasonable effectiveness of mathematics," as Nobel laureate physicist Eugene Wigner called it in 1960, in every step of my job. Whether I am struggling to understand which progenitor systems produce the stellar explosions known as type Ia supernovas or calculating the fate of Earth when our sun ultimately becomes a red giant, the tools I use and the models I develop are mathematical. The uncanny way that math captures the natural world has fascinated me throughout my career, and about 10 years ago I resolved to look into the issue more deeply.

At the core of this mystery lies an argument that mathematicians, physicists, philosophers, and cognitive scientists have had for centuries: Is math an invented set of tools, as Einstein believed? Or does it actually exist in some abstract realm, with humans merely discovering its truths? Many great mathematicians—including David Hilbert, Georg Cantor, and quite a few of the group known as Nicolas Bourbaki—have shared Einstein's view, associated with a school of thought called Formalism. But other illustrious thinkers—among them Godfrey Harold Hardy, Roger Penrose, and Kurt Gödel—have held the opposite view, Platonism.

This debate about the nature of mathematics rages on today and seems to elude an answer. I believe that by asking simply whether mathematics is invented or discovered, we ignore the possibility of a more intricate answer: both invention and discovery play a crucial role. I posit that together they account for why math works so well. Although eliminating the dichotomy between invention and discovery does not fully explain the unreasonable effectiveness of mathematics, the problem is so profound that even a partial step toward solving it is progress.

Invention and Discovery

Mathematics is unreasonably effective in two distinct ways, one I think of as active and the other as passive. Sometimes scientists create methods specifically for quantifying real-world phenomena. For example, Isaac Newton formulated calculus largely for the purpose of capturing motion and change, breaking them up into infinitesimally small frame-by-frame sequences. Of course, such active inventions are effective; the tools are, after all, made to order. What is surprising, however, is their stupendous accuracy in some cases. Take, for instance, quantum electrodynamics, the mathematical theory developed to describe how light and matter interact. When scientists use it to calculate the magnetic moment of the electron, the theoretical value agrees with the most recent experimental value—measured at 1.00115965218073 in the appropriate units in 2008—to within a few parts per trillion!

Even more astonishing, perhaps, mathematicians sometimes develop entire fields of study with no application in mind, and yet decades, even centuries, later physicists discover that these very branches make sense of their observations. Examples of this kind of passive effectiveness abound. French mathematician Évariste Galois, for example, developed group theory in the early 1800s for the sole purpose of determining the solvability of polynomial equations. Very broadly, groups are algebraic structures made up of sets of objects (say, the integers) united under some operation (for instance, addition) that obey specific rules (among them the existence of an identity element such as 0, which, when added to any integer, gives back that same integer). In 20th-century physics, this rather abstract field turned out to be the most fruitful way of categorizing elementary particles—the building blocks of matter. In the 1960s, physicists Murray Gell-Mann and Yuval Ne'eman independently showed that a specific group, referred to as SU(3), mirrored a behavior of subatomic particles called hadrons—a connection that ultimately laid the foundations for the modern theory of how atomic nuclei are held together.

The study of knots offers another beautiful example of passive effectiveness. Mathematical knots are similar to everyday knots, except that they have no loose ends. In the 1860s Lord Kelvin hoped to describe atoms as knotted tubes of ether. That misguided model failed to connect with reality, but mathematicians continued to analyze knots for many decades merely as an esoteric arm of pure mathematics. Amazingly, knot theory now provides important insights into string theory and loop quantum gravity—our current best attempts at articulating a theory of space-time that reconciles quantum mechanics with general relativity. Similarly, English mathematician Hardy's discoveries in number theory indirectly advanced the field of cryptography, despite Hardy's earlier proclamation that "no one has yet discovered any warlike purpose to be served by the theory of numbers." And in 1854 Bernhard Riemann described non-Euclidean geometries (previously formulated by Lovachevsky and Bolyai)—curious spaces in which it is possible to draw at least two parallel lines (or none at all) through a point not on a line. More than half a century later, Einstein invoked those geometries to build his general theory of relativity.

A pattern emerges: humans invent mathematical concepts by way of abstracting elements from the world around them—shapes, lines, sets, groups, and so forth—either for some specific purpose or simply for fun. They then go on to discover the connections among those concepts. Because this process of inventing and discovering is man-made—unlike the kind of discovery to which the Platonists subscribe—our mathematics is ultimately based on our perceptions and the mental pictures we can conjure. For instance, we possess an innate talent, called *subitizing*, for instantly recognizing quantity, which undoubtedly led to the concept of number. We are good at perceiving the edges of individual objects and at distinguishing between straight and curved lines and between different shapes, such as circles and el-lipses—abilities that probably led to the development of arithmetic and geometry. So, too, the repeated human experience of cause and effect at least partially contributed to the creation of logic and, with it, the notion that certain statements imply the validity of others.

Selection and Evolution

Michael Atiyah, one of the greatest mathematicians of the 20th cen-tury, has presented an elegant thought experiment that reveals just how perception colors which mathematical concepts we embrace—even ones as seemingly fundamental as numbers. German mathematician Leopold Kronecker famously declared, "God created the natural num-bers, all else is the work of man." But imagine if the intelligence in our world resided not with humankind but rather with a singular, isolated jellyfish, floating deep in the Pacific Ocean. Everything in its experi-ence would be continuous, from the flow of the surrounding water to its fluctuating temperature and pressure. In such an environment, lack-ing individual objects or indeed anything discrete, would the concept of number arise? If there were nothing to count, would numbers exist?

Like the jellyfish, we adopt mathematical tools that apply to our world—a fact that has undoubtedly contributed to the perceived ef-fectiveness of mathematics. Scientists do not choose analytical methods arbitrarily but rather on the basis of how well they predict the results of their experiments. When a tennis ball machine shoots out balls, you can use the natural numbers 1, 2, 3, and so on, to describe the flux of balls. When firefighters use a hose, however, they must invoke other concepts, such as volume or weight, to render a meaningful descrip-tion of the stream. So, too, when distinct subatomic particles collide in a particle accelerator, physicists turn to measures such as energy and momentum and not to the end number of particles, which would reveal only partial information about how the original particles collided be-cause additional particles can be created in the process.

Over time, only the best models survive. Failed models—such as French philosopher René Descartes's attempt to describe the motion of the planets by vortices of cosmic matter—die in their infancy. In con-trast, successful models evolve as new information becomes available.

For instance, very accurate measurements of the precession of the planet Mercury necessitated an overhaul of Newton's theory of gravity in the form of Einstein's general relativity. All successful mathematical concepts have a long shelf life: The formula for the surface area of a sphere remains as correct today as it was when Archimedes proved it around 250 BC. As a result, scientists of any era can search through a vast arsenal of formalisms to find the most appropriate methods.

Not only do scientists cherry-pick solutions, they also tend to select problems that are amenable to mathematical treatment. There exists, however, a whole host of phenomena for which no accurate mathematical predictions are possible, sometimes not even in principle. In economics, for example, many variables—the detailed psychology of the masses, to name one—do not easily lend themselves to quantitative analysis. The predictive value of any theory relies on the constancy of the underlying relations among variables. Our analyses also fail to fully capture systems that develop chaos, in which the tiniest change in the initial conditions may produce entirely different end results, prohibiting any long-term predictions. Mathematicians have developed statistics and probability to deal with such shortcomings, but mathematics itself is limited, as Austrian logician Gödel famously proved.

Symmetry of Nature

This careful selection of problems and solutions only partially accounts for the success of mathematics in describing the laws of nature. Such laws must exist in the first place! Luckily for mathematicians and physicists alike, universal laws appear to govern our cosmos: An atom 12 billion light-years away behaves just like an atom on Earth; light in the distant past and light today share the same traits; and the same gravitational forces that shaped the universe's initial structures hold sway over present-day galaxies. Mathematicians and physicists have invented the concept of symmetry to describe this kind of immunity to change.

The laws of physics seem to display symmetry with respect to space and time: They do not depend on where, from which angle, or when we examine them. They are also identical to all observers, irrespective of whether these observers are at rest, moving at constant speeds, or accelerating. Consequently, the same laws explain our results, whether the experiments occur in China, Alabama, or the Andromeda galaxy—and whether we conduct our experiment today or someone else does a billion years from now. If the universe did not possess these symmetries, any attempt to decipher nature's grand design—any mathematical model built on our observations—would be doomed because we would have to continuously repeat experiments at every point in space and time.

Even more subtle symmetries, called gauge symmetries, prevail within the laws that describe the subatomic world. For instance, because of the fuzziness of the quantum realm, a given particle can be a negatively charged electron or an electrically neutral neutrino, or a mixture of both—until we measure the electric charge that distinguishes between the two. As it turns out, the laws of nature take the same form when we interchange electrons for neutrinos or any mix of the two. The same holds true for interchanges of other fundamental particles. Without such gauge symmetries, it would have been difficult to provide a theory of the fundamental workings of the cosmos. We would be similarly stuck without locality—the fact that objects in our universe are influenced directly only by their immediate surroundings rather than by distant phenomena. Thanks to locality, we can attempt to assemble a mathematical model of the universe much as we might put together a jigsaw puzzle, starting with a description of the most basic forces among elementary particles and then building on additional pieces of knowledge.

Our current best mathematical attempt at unifying all interactions calls for yet another symmetry, known as supersymmetry. In a universe based on supersymmetry, every known particle must have an as-yet undiscovered partner. If such partners are discovered (for instance, once the Large Hadron Collider at CERN near Geneva reaches its full energy), it will be yet another triumph for the effectiveness of mathematics.

I started with two basic, interrelated questions: Is mathematics invented or discovered? And what gives mathematics its explanatory and predictive powers? I believe that we know the answer to the first question: Mathematics is an intricate fusion of inventions and discoveries. Concepts are generally invented, and even though all the correct relations among them existed before their discovery, humans still chose which ones to study. The second question turns out to be even more complex. There is no doubt that the selection of topics we address mathematically has played an important role in math's perceived effectiveness. But mathematics would not work at all were there no universal features to be discovered. You may now ask: Why are there universal laws of nature at all? Or equivalently: Why is our universe governed by certain symmetries and by locality? I truly do not know the answers, except to note that perhaps in a universe without these properties, complexity and life would have never emerged, and we would not be here to ask the question.

More to Explore

The Unreasonable Effectiveness of Mathematics in the Natural Sciences. Eugene Wigner in *Communications in Pure and Applied Mathematics*, Vol. 13, No. 1, pages 1–14; February 1960.

Pi in the Sky: Counting, Thinking, and Being. John D. Barrow. Back Bay Books, 1992.

Creation v. Discovery. Michael Atiyah in *Times Higher Education Supplement*, September 29, 1995.

Is God a Mathematician? Mario Livio. Simon & Schuster, 2009.

Scientific American Online, review of *Is God a Mathematician?* 2009.

Is Mathematics Discovered or Invented?

Timothy Gowers

The title of this chapter is a famous question. Indeed, perhaps it is a little too famous: It has been asked over and over again, and it is not clear what would constitute a satisfactory answer. However, I was asked to address it during the discussions that led to this volume, and since most of the participants in those discussions were not research mathematicians, I was in particular asked to give a mathematician's perspective on it.

One reason for the appeal of the question seems to be that people can use it to support their philosophical views. If mathematics is discovered, then it would appear that there is something out there that mathematicians are discovering, which in turn would appear to lend support to a Platonist conception of mathematics, whereas if it is invented, then that might seem to be an argument in favor of a nonrealist view of mathematical objects and mathematical truth.

But before a conclusion like that can be drawn, the argument needs to be fleshed out in detail. First, one must be clear what it means to say that some piece of mathematics has been discovered, and then one must explain, using that meaning, why a Platonist conclusion follows. I do not myself believe that this program can be carried out, but one can at least make a start on it by trying to explain the incontestable fact that almost all mathematicians who successfully prove theorems feel as though they are making discoveries. It is possible to think about this question in a nonphilosophical way, which is what I shall try to do. For instance, I shall consider whether there is an identifiable distinction between parts of mathematics that feel like discoveries and parts that feel like inventions. This question is partly psychological and partly a question about whether there are objective properties of mathematical statements that explain how they are perceived. The argument in favour of Platonism only needs *some* of mathematics to be discovered: If it turns out that there are two broad kinds of mathematics, then perhaps one can understand the distinction and formulate more precisely what mathematical discovery (as opposed to the mere producing of mathematics) is.

As the etymology of the word "discover" suggests, we normally talk of discovery when we find something that was, unbeknownst to us, already there. For example, Columbus is said to have discovered America (even if one can question that statement for other reasons), and Tutankhamun's tomb was discovered by Howard Carter in 1922. We say this even when we cannot directly observe what has been discovered: For instance, J. J. Thompson is famous as the discoverer of the electron. Of greater relevance to mathematics is the discovery of facts: We discover *that* something is the case. For example, it would make perfectly good sense to say that Bernstein and Woodward discovered (or contributed to the discovery) that Nixon was linked to the Watergate burglary.

In all these cases, we have some phenomenon, or fact, that is brought to our attention by the discovery. So one might ask whether this transition from unknown to known could serve as a definition of discovery. But a few examples show that there is a little more to it than that. For instance, an amusing fact, known to people who like doing cryptic crosswords, is that the words "carthorse" and "orchestra" are anagrams. I presume that somebody somewhere was the first person to notice this fact, but I am inclined to call it an observation (hence my use of the word "notice") rather than a discovery. Why is this? Perhaps it is because the words "carthorse" and "orchestra" were there under our noses all the time and what has been spotted is a simple relationship between them. But why could we not say that the relationship is discovered even if the words were familiar? Another possible explanation is that once the relationship is pointed out, one can easily verify that it holds: You don't have to travel to America or Egypt, or do a delicate scientific experiment, or get access to secret documents.

As far as evidence for Platonism is concerned, the distinction between discovery and observation is not especially important: If you notice something, then that something must have been there for you to notice, just as if you discover it, then it must have been there for you to discover. So let us think of observation as a mild kind of discovery rather than as a fundamentally different phenomenon.

How about invention? What kinds of things do we invent? Machines are an obvious example: We talk of the invention of the steam engine, or the airplane, or the mobile phone. We also invent games: For instance, the British invented cricket—and more to the point, that is an appropriate way of saying what happened. Art supplies us with a more interesting example. One would never talk of a single work of art being invented, but it does seem to be possible to invent a style or a technique. For example, Picasso did not invent *Les Desmoiselles d'Avignon*, but he and Braque are credited with inventing cubism.

A common theme that emerges from these examples is that what we invent tends not to be individual objects: Rather, we invent general *methods* for producing objects. When we talk of the invention of the steam engine, we are not talking about one particular instance of steam-enginehood, but rather of the idea—that a clever arrangement of steam, pistons, etc., can be used to drive machines—that led to the building of many steam engines. Similarly, cricket is a set of rules that has led to many games of cricket, and cubism is a general idea that led to the painting of many cubist pictures.

If somebody wants to argue that the fact of mathematical discovery is evidence for a Platonist view of mathematics, then what they will be trying to show is that certain abstract entities have an independent existence, and certain facts about those entities are true for much the same sort of reason that certain facts about concrete entities are true. For instance, the statement "There are infinitely many prime numbers" is true, according to this view, because there really are infinitely many natural numbers out there, and it really is the case that infinitely many of them are prime.

A small remark one could make here is that it is also possible to use the concept of invention as an argument in favor of an independent existence for abstract concepts. Indeed, our examples of invention all involve abstraction in a crucial way: The steam engine, as we have just noted, is an abstract concept, as are the rules of cricket. Cubism is a more problematic example, as it is less precisely defined, but it is undoubtedly abstract rather than concrete. Why do we not say that these abstract concepts are brought into existence when we invent them?

One reason is that we feel that independently existing abstract concepts should be timeless. So we do not like the idea that when the British invented the rules of cricket, they reached out into the abstract realm and brought the rules into existence. A more appealing picture would be that they selected the rules of cricket from a vast "rule space" that consists of all possible sets of rules (most of which give rise to terrible games). A drawback with this second picture is that it fills up the abstract realm with a great deal of junk, but perhaps it really is like that. For example, it is supposed to contain all the real numbers, all but countably many of which are undefinable.

Another argument against the idea that one brings an abstract concept into existence when one invents it is that the concepts that we invent are not fundamental enough: They tend to be methods for dealing with other objects, either abstract or concrete, that are much simpler. For example, the rules of cricket describe constraints on a set of procedures that are carried out by 22 players, a ball, and two wickets. From an ontological point of view, the players, ball, and wickets seem more secure than the constraints on how they behave.

Earlier, I commented that we do not normally talk of inventing a single work of art. However, we do not discover it either. A commonly used word for what we do is "create." And most people, if asked, would say that this kind of creation has more in common with invention than with discovery, just as observation has more in common with discovery than with invention.

Why is this? Well, in both cases what is brought into existence has many arbitrary features: If we could turn the clock back to just before cricket was invented and run the world all over again, it is likely that we would see the invention of a similar game, but unlikely that its rules would be identical to those of the actual game of cricket. (One might object that if the laws of physics are deterministic, then the world would develop precisely as it did the first time. In that case, one could make a few small random changes before the rerun.) Similarly, if somebody had accidentally destroyed *Les Desmoiselles d'Avignon* just after Picasso started work on it, forcing him to start again, it is likely that he would have produced a similar but perceptibly different painting. By contrast, if Columbus had not existed, then somebody else would have discovered *America* and not just some huge landmass of a broadly similar kind on the other side of the Atlantic. And the fact that "carthorse" and "orchestra" are anagrams is independent of who was the first to observe it.

With these thoughts in mind, let us turn to mathematics. Again, it will help to look at some examples of what people typically say about various famous parts of the subject. Let me list some discoveries, some observations, and some inventions. (I cannot think of circumstances where I would definitely want to say that a piece of mathematics was created.) Later I will try to justify why each item is described in the way it is.

A few well-known discoveries are the formula for the quadratic equation, the absence of a similar formula for the quintic, the monster group, and the fact that there are infinitely many primes. A few observations are that the number of primes less than 100 is 25, that the last digits of the powers of 3 form the sequence 3, 9, 7, 1, 3, 9, 7,1, . . . , and that the number 10,001 factors as 73 times 137. An intermediate case is the fact that if you define an infinite sequence $z_0, z_1, z_2, \ldots$ of complex numbers by setting $z_0 = 0$ and $z_n = z_{n-1}^2 + C$ for every $n > 0$, then the set of all complex numbers C for which the sequence does not tend to infinity, now called the Mandelbrot set, has a remarkably complicated structure. (I regard this as intermediate because, although Mandelbrot and others stumbled on it almost by accident, it has turned out to be an object of fundamental importance in the theory of dynamical systems.)

On the other side, it is often said that Newton and Leibniz independently invented calculus. (I planned to include this example, and was heartened when, quite by coincidence, on the day that I am writing this paragraph, there was a plug for a radio program about their priority dispute, and the word "invented" was indeed used.) One also sometimes talks of mathematical theories (as opposed to theorems) being invented: It does not sound ridiculous to say that Grothendieck invented the theory of schemes, though one might equally well say "introduced" or "developed." Similarly, any of these three words would be appropriate for describing what Cohen did to the method of forcing, which he used to prove the independence of the continuum hypothesis. From our point of view, what is interesting is that the words "invent," "introduce," and "develop" all carry with them the suggestion that some general technique is brought into being.

A mathematical object about which there might be some dispute is the number i, or more generally the complex number system. Were complex numbers discovered or invented? Or rather, would mathematicians normally refer to the arrival of complex numbers into mathematics using a discovery-type word or an invention-type word? If you type the phrases "complex numbers were invented" and "complex numbers were discovered" into Google, you get approximately the same number of hits (between 4,500 and 5,000 in both cases), so there appears to be no clear answer. But this too is a useful piece of data. A similar example is non-Euclidean geometry, though here "discovery of non-Euclidean geometry" outnumbers "invention of non-Euclidean geometry" by a ratio of about 3 to 1.

Another case that is not clear-cut is that of *proofs*: Are they discovered or invented? Sometimes a proof seems so natural—mathematicians often talk of "the right proof" of a statement, meaning not that it is the only correct proof but that it is the one proof that truly explains why the statement is true—that the word "discover" is the obvious word to use. But sometimes it feels more appropriate to say something like, "Conjecture 2.5 was first proved in 1990, but in 2002 Smith came up with an ingenious and surprisingly short argument that actually establishes a slightly more general result." One could say "discovered" instead of "came up with" in that sentence, but the latter captures better the idea that Smith's argument was just one of many that there might have been and that Smith did not simply stumble on it by accident.

Let us take stock at this point, and see whether we can explain what it is about a piece of mathematics that causes us to put it into one of the three categories: discovered, invented, or not clearly either.

The nonmathematical examples suggest that discoveries and observations are usually of objects or facts over which the discoverer has no control, whereas inventions and creations are of objects or procedures with many features that could be chosen by the inventor or creator. We also drew some more refined, but less important, distinctions within each class. A discovery tends to be more notable than an observation and less easy to verify afterward. And inventions tend to be more general than creations.

Do these distinctions continue to hold in much the same form when we come to talk about mathematics? I claimed earlier that the formula for the quadratic was discovered, and when I try out the phrase "the invention of the formula for the quadratic," I find that I do not like it, for exactly the reason that the solutions of $ax^2 + bx + c$ *are* the numbers $(-b \pm \sqrt{b^2 - 4ac})/2a$. Whoever first derived that formula did not have any choice about what the formula would eventually be. It is, of course, possible to notate the formula differently, but that is another matter. I do not want to get bogged down in a discussion of what it means for two formulas to be "essentially the same," so let me simply say that the formula itself was a discovery but that different people have *come up with* different ways of expressing it. However, this kind of concern will reappear when we look at other examples.

The insolubility of the quintic is another straightforward example. *It is* insoluble by radicals, and nothing Abel did could have changed that. So his famous theorem was a discovery. However, aspects of his *proof* would be regarded as invention—there have subsequently been different looking proofs. This notion is particularly clear with the closely related work of Galois, who is credited with the invention of group theory. (The phrase "invention of group theory" has 40,300 entries in Google, compared with 10 for "discovery of group theory.")

The monster group is a more interesting case. It first entered the mathematical scene when Fischer and Griess predicted its existence in 1973. But what does that mean? If they could refer to the monster group at all, then does that not imply that it existed? The answer is simple: They predicted that a group with certain remarkable *properties* (one of which is its huge size—hence the name) existed and was unique. So to say "I believe that the monster group exists" was shorthand for "I believe that there exists a group with these amazing properties," and the name "monster group" was referring to a hypothetical entity.

The existence and uniqueness of the monster group were indeed proved, though not until 1982 and 1990, respectively, and it is not quite clear whether we should regard this mathematical advance as a discovery or an invention. If we ignore the story and condense 17 years to an instant, then it is tempting to say that the monster group was there all along until it was discovered by group theorists. Perhaps one could even add a little detail: Back in 1973, people started to have reason to suppose that it existed, and they finally bumped into it in 1982.

But how did this "bumping" take place? Griess did not prove in some indirect way that the monster group had to exist (though such proofs are possible in mathematics). Rather, he *constructed* the group. Here, I am using the word that all mathematicians would use. To construct it, he constructed an auxiliary object, a complicated algebraic structure now known as the Griess algebra, and showed that the symmetries of this algebra formed a group with the desired properties. However, this method is not the only way of obtaining the monster group: There are other constructions that give rise to groups that have the same properties, and hence, by the uniqueness result, are isomorphic to it. So it seems that Griess had some control over the process by which he built the monster group, even if what he ended up building was determined in advance. Interestingly, the phrase "construction of the monster group" is much more popular on Google than the phrase "discovery of the monster group" (8,290 to 9), but if you change it to "the construction of the monster group," then it becomes much less popular (6 entries), reflecting the fact that there are many different constructions.

Another question one might ask is this. If we do decide to talk about the discovery of the monster group, are we talking about the discovery of an *object*, the monster group, or of a *fact*, the fact that there exists a group with certain properties and that that group is unique? Certainly, the second is a better description of the work that the group theorists involved actually did, and the word "construct" is a better word than "discover" at describing how they proved the existence part of this statement.

The other discoveries and observations listed earlier appear to be more straightforward, so let us turn to the examples on the invention side.

A straightforward use of the word "invention" in mathematics is to refer to the way general theories and techniques come into being. This way of coming into being certainly covers the example of calculus, which is not an object, or a single fact, but rather a large collection of facts and methods that greatly increase your mathematical power when you are familiar with them. It also covers Cohen's technique of forcing: Again, there are theorems involved, but what is truly interesting about forcing is that it is a general and adaptable method for proving independence statements in set theory.

I suggested earlier that inventors should have some control over what they invent. That applies to these examples: There is no clear criterion that says which mathematical statements are part of calculus, and there are many ways of presenting the theory of forcing (and, as I mentioned earlier, many generalizations, modifications, and extensions of Cohen's original ideas).

How about the complex number system? At first sight, this system does not look at all like an invention. After all, it is provably unique (up to the isomorphism that sends $a + bi$ to $a - bi$), and it is an object rather than a theory or a technique. So why do people sometimes call it an invention, or at the very least feel a little uneasy about calling it a discovery?

I do not have a complete answer to this question, but I suspect that the reason it is a somewhat difficult example is similar to the reason that the monster group is difficult, which is that one can "construct" the complex numbers in more than one way. One approach is to use something like the way they were constructed historically (my knowledge of the history is patchy, so I shall not say *how* close the resemblance is). One simply introduces a new symbol, i, and declares that it behaves much like a real number, obeying all the usual algebraic rules, and has the additional property that $i^2 = -1$. From this setup, one can deduce that

$$(a + bi)(c + di) = ac + bci + adi + bdi^2 = (ad - bd) + (ad + bc)i$$

and many other facts that can be used to build up the theory of complex numbers. A second approach, which was introduced much later to demonstrate that the complex number system was consistent if the real number system was, is to define a complex number to be an ordered pair (a, b) of real numbers, and to stipulate that addition and multiplication of these ordered pairs are given by the following rules:

$$(a, b) + (c, d) = (a + c, b + d)$$
$$(a, b) + (c, d) = (ac - bd, ad + bc)$$

This second method is often used in university courses that build up the number systems rigorously. One proves that these ordered pairs form a field under the two given operations, and finally one says, "From now on I shall write $a + bi$ instead of (a, b)."

Another reason for our ambivalence about the complex numbers is that they feel less real than real numbers. (Of course, the names given to these numbers reflect this notion rather unsubtly.) We can directly relate the real numbers to quantities such as time, mass, length, temperature, and so on (though for this usage, we never need the infinite precision of the real number system), so it feels as though they have an independent existence that we observe. But we do not run into the complex numbers in that way. Rather, we play what feels like a sort of game—imagine what would happen if -1 *did* have a square root.

But why in that case do we not feel happy just to say that the complex numbers were invented? The reason is that the game is much more interesting than we had any right to expect, and it has had a huge influence even on those parts of mathematics that are about real numbers or even integers. It is as though after our one small act of inventing i, the game took over and we lost control of the consequences. (Another example of this phenomenon is Conway's famous game of Life. He devised a few simple rules, by a process that one would surely want to regard as closer to invention than discovery. But once he had done so, he found that he had created a world full of unexpected phenomena that he had not put there, so to speak. Indeed, most of them were discovered—to use the obvious word—by other people.)

Why is "discovery of non-Euclidean geometry" more popular than "invention of non-Euclidean geometry"? This is an interesting case because there are two approaches to the subject, one axiomatic and one concrete. One could talk about non-Euclidean geometry as the discovery of the remarkable fact that a different set of axioms, where the parallel postulate is replaced by a statement that allows a line to have several parallels through any given point, is consistent. Alternatively, one could think of it as the construction of models in which those axioms are true. Strictly speaking, one needs the second for the first, but if one explores in detail the consequences of the axioms and proves all sorts of interesting theorems without ever reaching a contradiction, that can be quite impressive evidence for their consistency. It is probably because the consistency interests us more than the particular choice of model, combined with the fact that any two models of the hyperbolic plane are isometric, that we usually call it a discovery. However, Euclidean geometry (wrongly) feels more "real" than hyperbolic geometry, and there is no single model of hyperbolic geometry that stands out as the most natural one; these two facts may explain why the word "invention" is sometimes used.

My final example was that of proofs, which I claimed could be discovered *or* invented, depending on the nature of the proof. Of course, these are by no means the only two words or phrases that one might use: Some others are "thought of," "found," and "came up with." Often one regards the proof less as an object than as a process and focuses on what is proved, as is shown in sentences such as, "After a long struggle, they eventually managed to prove or establish or show or demonstrate that . . ." Proofs illustrate once again the general point that we use discovery words when the author has less control and invention words when there are many choices to be made. Where, one might ask, does the choice come from? This is a fascinating question in itself, but let me point out just one source of choice and arbitrariness: Often a proof requires one to show that a certain mathematical object or structure exists (either as the main statement or as some intermediate lemma), and often the object or structure in question is far from unique.

Before drawing any conclusions from these examples, I would like to discuss briefly another aspect of the question. I have been looking at it mainly from a linguistic point of view, but, as I mentioned right at the beginning, it also has a strong psychological component: When one is doing mathematical research, it sometimes feels more like discovery and sometimes more like invention. What is the difference between the two experiences?

Since I am more familiar with myself than with anybody else, let me draw on my own experience. In the mid-1990s, I started on a research project that has occupied me in one way or another ever since. I was thinking about a theorem that I felt ought to have a simpler proof than the two that were then known. Eventually, I found one (here I am using the word that comes naturally); unfortunately it was not simpler, but it gave important new information. The process of finding this proof felt much more like discovery than invention because by the time I reached the end, the structure of the argument included many elements that I had not even begun to envisage when I started working on it. Moreover, it became clear that there was a large body of closely related facts that added up to a coherent and yet-to-be-discovered theory. (At this stage, they were not proved facts, and not always even precisely stated facts. It was just clear that "something was going on" that needed to be investigated.) I and several others have been working to develop this theory, and theorems have been proved that would not even have been stated as conjectures 15 years ago.

Why did this work feel like discovery rather than invention? Once again, it is connected with control: I was not selecting the facts I happened to like from a vast range of possibilities. Rather, certain statements stood out as obviously natural and important. Now that the theory is more developed, it is less clear which facts are central and which more peripheral, and for that reason, the enterprise feels as though it has an invention component as well.

A few years earlier, I had a different experience: I found a counterexample to an old conjecture in the theory of Banach spaces. To do this, I constructed a complicated Banach space. This construction felt partly like an invention—I did have arbitrary choices, and many other counterexamples have subsequently been found—and partly like a discovery—much of what I did was in response to the requirements of the problem and felt like the natural thing to do, and a similar example was discovered independently by someone else (and even the later examples use similar techniques). So this is another complicated situation to analyze, but the reason it is complicated is simply that the question of how much control I had is a complicated one.

What conclusion should we draw from all these examples and from how we naturally seem to regard them? First, it is clear that the question with which we began is rather artificial. For a start, the idea that either all of mathematics is discovered or all of mathematics is invented is ridiculous. But even if we look at the origins of individual pieces of mathematics, we are not forced to use the word "discover" or "invent," and we often don't.

Nevertheless, there does seem to be a spectrum of possibilities, with some parts of mathematics feeling more like discoveries and others more like inventions. It is not always easy to say which are which, but there does seem to be one feature that correlates strongly with whether we prefer to use a discovery-type word or an invention-type word. That feature is the control that we have over what is produced. This feature, as I have argued, even helps to explain why the doubtful cases are doubtful.

If this difference is correct (perhaps after some refinement), what philosophical consequences can we draw from it? I suggested at the beginning that the answer to the question did not have any bearing on questions such as "Do numbers exist?" or "Are mathematical statements true because the objects they mention really do relate to each other in the ways described?" My reason for that suggestion is that pieces of mathematics have objective features that explain how much control we have over them. For instance, as I mentioned earlier, the proof of an existential statement may well be far from unique, for the simple reason that there may be many objects with the required properties. But this statement is a straightforward mathematical phenomenon. One could accept my analysis and believe that the objects in question "really exist," or one could view the statements that they exist as moves in games played with marks on paper, or one could regard the objects as convenient fictions. The fact that some parts of mathematics are unexpected and others not, that some solutions are unique and others multiple, that some proofs are obvious and others take a huge amount of work to produce—all these have a bearing on how we describe the process of mathematical production, and all of them are entirely independent of one's philosophical position.

The Unplanned Impact of Mathematics

PETER ROWLETT

As a child, I read a joke about someone who invented the electric plug and had to wait for the invention of a socket to put it in. Who would invent something so useful without knowing what purpose it would serve? Mathematics often displays this astonishing quality. Trying to solve real-world problems, researchers often discover that the tools they need were developed years, decades, or even centuries earlier by mathematicians with no prospect of, or care for, applicability. And the toolbox is vast, because once a mathematical result is proven to the satisfaction of the discipline, it doesn't need to be reevaluated in the light of new evidence or refuted, unless it contains a mistake. If it was true for Archimedes, then it is true today.

The mathematician develops topics that no one else can see any point in pursuing or pushes ideas far into the abstract, well beyond where others would stop. Chatting with a colleague over tea about a set of problems that ask for the minimum number of stationary guards needed to keep under observation every point in an art gallery, I outlined the basic mathematics, noting that it only works on a two-dimensional floor plan and breaks down in three-dimensional situations, such as when the art gallery contains a mezzanine. "Ah," he said, "but if we move to 5D we can adapt . . ." This extension and abstraction without apparent direction or purpose is fundamental to the discipline. Applicability is not the reason we work, and plenty that is not applicable contributes to the beauty and magnificence of our subject.

There has been pressure in recent years for researchers to predict the impact of their work before it is undertaken. Alan Thorpe, then chair of Research Councils UK, was quoted by *Times Higher Education* (22 October 2009) as saying, "We have to demonstrate to the taxpayer that this is an investment, and we do want researchers to think about what the impact of their work will be." The U.S. National Science Foundation is similarly focused on broader impacts of research proposals (*Nature* **465**, 416–418; 2010). However, predicting impact is extremely problematic. The latest *International Review of Mathematical Sciences* (Engineering and Physical Sciences Research Council; 2010), an independent assessment of the quality and impact of U.K. research, warned that even the most theoretical mathematical ideas "can be useful or enlightening in unexpected ways, sometimes several decades after their appearance."

There is no way to guarantee in advance which pure mathematics will later find application. We can only let the process of curiosity and abstraction take place, let mathematicians obsessively take results to their logical extremes, leaving relevance far behind, and wait to see which topics turn out to be extremely useful. If not, when the challenges of the future arrive, we won't have the right piece of seemingly pointless mathematics at hand.

To illustrate this notion, I asked members of the British Society for the History of Mathematics (including myself) for unsung stories of the unplanned impact of mathematics (beyond the use of number theory in modern cryptography, or that the mathematics to operate a computer existed when one was built, or that imaginary numbers became essential to the complex calculations that fly airplanes). Here follow seven examples.

From Quaternions to Lara Croft

MARK MCCARTNEY AND TONY MANN
University of Ulster, Newtownabbey, UK;
University of Greenwich, London, UK

Famously, the idea of quaternions came to the Irish mathematician William Rowan Hamilton on 16 October 1843 as he was walking over Brougham Bridge, Dublin. He marked the moment by carving the equations into the stonework of the bridge. Hamilton had been seeking a way to extend the complex-number system into three dimensions: his insight on the bridge was that it was necessary instead to move to four dimensions to obtain a consistent number system. Whereas complex numbers take the form $a + ib$, where a and b are real numbers and i is the square root of -1, quaternions have the form $a + bi + cj + dk$, where the rules are $i^2 = j^2 = k^2 = ijk = -1$.

Hamilton spent the rest of his life promoting the use of quaternions, as mathematics both elegant in its own right and useful for solving problems in geometry, mechanics, and optics. After his death, the torch was carried by Peter Guthrie Tait (1831–1901), professor of natural philosophy at the University of Edinburgh. William Thomson (Lord Kelvin) wrote of Tait, "We have had a thirty-eight-year war over quaternions." Thomson agreed with Tait that they would use quaternions in their important joint book the *Treatise on Natural Philosophy* (1867) wherever they were useful. However, their complete absence from the final manuscript shows that Thomson was not persuaded of their value.

By the close of the nineteenth century, vector calculus had eclipsed quaternions, and mathematicians in the twentieth century generally followed Kelvin rather than Tait, regarding quaternions as a beautiful, but sadly impractical, historical footnote.

So it was a surprise when a colleague who teaches computer-games development asked which mathematics module students should take to learn about quaternions. It turns out that they are particularly valuable for calculations involving three-dimensional rotations, where they have various advantages over matrix methods. This fact makes them indispensable in robotics and computer vision, and in ever-faster graphics programming.

Tait would no doubt be happy to have finally won his "war" with Kelvin. And Hamilton's expectation that his discovery would be of great benefit has been realized, after 150 years, in gaming, an industry estimated to be worth more than US$100 billion worldwide.

From Geometry to the Big Bang

GRAHAM HOARE
Correspondence editor, Mathematics Today

In 1907, Albert Einstein's formulation of the equivalence principle was a key step in the development of the general theory of relativity. His idea, that the effects of acceleration are indistinguishable from the effects of a uniform gravitational field, depends on the equivalence between gravitational mass and inertial mass. Einstein's essential insight was that gravity manifests itself in the form of space-time curvature; gravity is no longer regarded as a force. How matter curves the surrounding space-time is expressed by Einstein's field equations. He published his general theory in 1915; its origins can be traced back to the middle of the previous century.

In his brilliant Habilitation lecture of 1854, Bernhard Riemann introduced the principal ideas of modern differential geometry—*n*-dimensional spaces, metrics and curvature, and the way in which curvature controls the geometric properties of space—by inventing the concept of a manifold. Manifolds are essentially generalizations of shapes, such as the surface of a sphere or a torus, on which one can do calculus. Riemann went far beyond the conceptual frameworks of Euclidean and non-Euclidean geometry. He foresaw that his manifolds could be models of the physical world.

The tools developed to apply Riemannian geometry to physics were initially the work of Gregorio Ricci-Curbastro, beginning in 1892 and extended with his student Tullio Levi-Civita. In 1912, Einstein enlisted the help of his friend, the mathematician Marcel Grossmann, to use this "tensor calculus" to articulate his deep physical insights in mathematical form. He employed Riemann manifolds in four dimensions: three for space and one for time (space-time).

It was the custom at the time to assume that the universe is static. But Einstein soon found that his field equations when applied to the whole universe did not have any static solutions. In 1917, to make a static universe possible, Einstein added the cosmological constant to his original field equations. Reasons for believing in an explosive origin to the universe, the Big Bang, were put forward by Aleksander Friedmann in his 1922 study of Einstein's field equations in a cosmological context. Grudgingly accepting the irrefutable evidence of the expansion of the universe, Einstein deleted the constant in 1931, referring to it as "the biggest blunder" of his life.

From Oranges to Modems

EDMUND HARRISS
University of Arkansas, Fayetteville, USA

In 1998, mathematics was suddenly in the news. Thomas Hales of the University of Pittsburgh, Pennsylvania, had proved the Kepler conjecture, showing that the way grocers stack oranges is the most efficient way to pack spheres. A problem that had been open since 1611 was finally solved! On the television a grocer said, "I think that it's a waste of time and taxpayers' money." I have been mentally arguing with that grocer ever since: Today the mathematics of sphere packing enables modern communication; it is at the heart of the study of channel coding and error-correction codes.

In 1611, Johannes Kepler suggested that the grocer's stacking was the most efficient, but he was not able to give a proof. It turned out to be a difficult problem. Even the simpler question of the best way to pack circles was only proved in 1940 by László Fejes Tóth. Also in the seventeenth century, Isaac Newton and David Gregory argued over the kissing problem: How many spheres can touch a given sphere with no overlaps? In two dimensions, it is easy to prove that the answer is 6. Newton thought that 12 was the maximum in three dimensions. It is, but only in 1953 did Kurt Schütte and Bartel van der Waerden give a proof.

The kissing number in four dimensions was proved to be 24 by Oleg Musin in 2003. In five dimensions, we can say only that it lies between 40 and 44. Yet we do know that the answer in eight dimensions is 240, proved back in 1979 by Andrew Odlyzko of the University of Minnesota, Minneapolis. The same paper had an even stranger result: The answer in 24 dimensions is 196,560. These proofs are simpler than the result for three dimensions, and they relate to two incredibly dense packings of spheres, called the E8 lattice in eight dimensions and the Leech lattice in 24 dimensions.

This figuring is all quite magical, but is it useful? In the 1960s, an engineer called Gordon Lang believed so. Lang was designing the systems for modems and was busy harvesting all the mathematics he could find.

He needed to send a signal over a noisy channel, such as a phone line. The natural way is to choose a collection of tones for signals. But the sound received may not be the same as the one sent. To solve this problem, he described the sounds by a list of numbers. It was then simple to find which of the signals that might have been sent was closest to the signal received. The signals can then be considered as spheres, with wiggle room for noise. To maximize the information that can be sent, these "spheres" must be packed as tightly as possible.

In the 1970s, Lang developed a modem with eight-dimensional signals, using E8 packing. This solution helped to open up the Internet because data could be sent over the phone, instead of relying on specifically designed cables. Not everyone was thrilled. Donald Coxeter, who had helped Lang understand the mathematics, said he was "appalled that his beautiful theories had been sullied in this way."

From Paradox to Pandemics

JUAN PARRONDO AND NOEL-ANN BRADSHAW
University of Madrid, Spain; University of Greenwich, London, UK

In 1992, two physicists proposed a simple device to turn thermal fluctuations at the molecular level into directed motion: a "Brownian ratchet." It consists of a particle in a flashing asymmetric field. Switching the field on and off induces the directed motion, explained Armand Ajdari of the School of Industrial Physics and Chemistry in Paris and Jacques Prost of the Curie Institute in Paris.

Parrondo's paradox, discovered in 1996 by one of us (J.P.), captures the essence of this phenomenon mathematically, translating it into a simpler and broader language: gambling games. In the paradox, a gambler alternates between two games, both of which lead to an expected loss in the long term. Surprisingly, by switching between them, one can produce a game in which the expected outcome is positive. The term "Parrondo effect" is now used to refer to an outcome of two combined events being very different from the outcomes of the individual events.

A number of applications of the Parrondo effect are now being investigated in which chaotic dynamics can combine to yield nonchaotic behavior. For example, the effect can be used to model the population dynamics in outbreaks of viral diseases and offers prospects of reducing the risks of share-price volatility. Also, it plays a leading part in the plot of Richard Armstrong's 2006 novel, *God Doesn't Shoot Craps: A Divine Comedy*.

From Gamblers to Actuaries

PETER ROWLETT
University of Birmingham, UK

In the sixteenth century, Girolamo Cardano was a mathematician and a compulsive gambler. Tragically for him, he squandered most of the money he inherited and earned. Fortunately for modern actuarial science, he wrote in the mid-1500s what is considered to be the first work in modern probability theory, *Liber de ludo aleae*, finally published in a collection in 1663.

Around a century after the creation of this theory, another gambler, Chevalier de Méré, had a dilemma. He had been offering a game in which he bet he could throw a six in four rolls of a die, and had done well out of it. He varied the game in a way that seemed sensible, betting he could throw a double six with two dice in 24 rolls. He had calculated the chances of winning in both games as equivalent but found that he lost money in the long run playing the second game. Confused, he asked his friend Blaise Pascal for an explanation. Pascal wrote to Pierre de Fermat in 1654. The ensuing correspondence laid the foundations for probability theory, and when Christiaan Huygens learned of the results, he wrote the first published work on probability, *De Ratiociniis in Ludo Aleae* (published in 1657).

In the late seventeenth century, Jakob Bernoulli recognized that probability theory could be applied much more widely than to games of chance. He wrote *Ars Conjectandi* (published, after his death, in 1713), which consolidated and extended the probability work by Cardano, Fermat, Pascal, and Huygens. Bernoulli built on Cardano's discovery that with sufficient rolls of a fair, six-sided die we can expect each outcome to appear around one-sixth of the time, but that if we roll one die six times we shouldn't expect to see each outcome precisely once. Bernoulli gave a proof of the law of large numbers, which says that the larger a sample, the more closely the sample characteristics match those of the parent population.

Insurance companies had been limiting the number of policies they sold. As policies are based on probabilities, each policy sold seemed to incur an additional risk, the cumulative effect of which, it was feared, could ruin a company. Beginning in the eighteenth century, companies began their current practice of selling as many policies as possible, because, as Bernoulli's law of large numbers showed, the bigger the volume, the more likely their predictions are to be accurate.

From Bridges to DNA

JULIA COLLINS
University of Edinburgh, UK

When Leonhard Euler proved to the people of Königsberg in 1735 that they could not traverse all of their seven bridges in one trip, he invented a new kind of mathematics: one in which distances didn't matter. His solution relied only on knowing the relative arrangements of the bridges, not on how long they were or how big the land masses were. In 1847,

Johann Benedict Listing finally coined the term "topology" to describe this new field, and for the next 150 years or so, mathematicians worked to understand the implications of its axioms.

For most of that time, topology was pursued as an intellectual challenge, with no expectation of it being useful. After all, in real life, shape and measurement are important: A doughnut is not the same as a coffee cup. Who would ever care about five-dimensional holes in abstract 11-dimensional spaces, or whether surfaces had one or two sides? Even practical-sounding parts of topology, such as knot theory, which had its origins in attempts to understand the structure of atoms, were thought to be useless for most of the nineteenth and twentieth centuries.

Suddenly, in the 1990s, applications of topology started to appear— slowly at first, but gaining momentum until now it seems as if there are few areas in which topology is not used. Biologists learn knot theory to understand DNA. Computer scientists are using braids—intertwined strands of material running in the same direction—to build quantum computers, while colleagues down the corridor use the same theory to get robots moving. Engineers use one-sided Möbius strips to make more efficient conveyer belts. Doctors depend on homology theory to do brain scans, and cosmologists use it to understand how galaxies form. Mobile-phone companies use topology to identify the holes in network coverage; the phones themselves use topology to analyze the photos they take.

It is precisely because topology is free of distance measurements that it is so powerful. The same theorems apply to any knotted DNA, regardless of how long it is or what animal it comes from. We don't need different brain scanners for people with different-sized brains. When global positioning system data about mobile phones are unreliable, topology can still guarantee that those phones receive a signal. Quantum computing won't work unless we can build a robust system impervious to noise, so braids are perfect for storing information because they don't change if you wiggle them. Where will topology turn up next?

From Strings to Nuclear Power

Chris Linton
Loughborough University, UK

Series of sine and cosine functions were used by Leonhard Euler and others in the eighteenth century to solve problems, notably in the study of vibrating strings and in celestial mechanics. But it was Joseph Fourier, at the beginning of the nineteenth century, who recognized the great practical utility of these series in heat conduction and began to develop a general theory. Thereafter, the list of areas in which Fourier series were found to be useful grew rapidly to include acoustics, optics, and electric circuits. Nowadays, Fourier methods underpin large parts of science and engineering and many modern computational techniques.

However, the mathematics of the early nineteenth century was inadequate for the development of Fourier's ideas, and the resolution of the numerous problems that arose challenged many of the great minds of the time. This in turn led to new mathematics. For example, in the 1830s, Gustav Lejeune Dirichlet gave the first clear and useful definition of a function, and Bernhard Riemann in the 1850s and Henri Lebesgue in the 1900s created rigorous theories of integration. What it means for an infinite series to converge turned out to be a particularly slippery animal, but it was gradually tamed by theorists such as Augustin-Louis Cauchy and Karl Weierstrass, working in the 1820s and 1850s, respectively. In the 1870s, Georg Cantor's first steps toward an abstract theory of sets came about through analyzing how two functions with the same Fourier series could differ.

The crowning achievement of this mathematical trajectory, formulated in the first decade of the twentieth century, is the concept of a Hilbert space. Named after the German mathematician David Hilbert, this is a set of elements that can be added and multiplied according to a precise set of rules, with special properties that allow many of the tricky questions posed by Fourier series to be answered. Here the power of mathematics lies in the level of abstraction, and we seem to have left the real world behind.

Then in the 1920s, Hermann Weyl, Paul Dirac, and John von Neumann recognized that this concept was the bedrock of quantum mechanics, since the possible states of a quantum system turn out to be elements of just such a Hilbert space. Arguably, quantum mechanics is the most successful scientific theory of all time. Without it, much of our modern technology—lasers, computers, flat-screen televisions, nuclear power—would not exist.

An Adventure in the Nth Dimension

Brian Hayes

The area enclosed by a circle is πr^2. The volume inside a sphere is $\frac{4}{3}\pi r^3$. These are formulas I learned too early in life. Having committed them to memory as a schoolboy, I ceased to ask questions about their origin or meaning. In particular, it never occurred to me to wonder how the two formulas are related, or whether they could be extended beyond the familiar world of two- and three-dimensional objects to the geometry of higher-dimensional spaces. What's the volume bounded by a four-dimensional sphere? Is there some master formula that gives the measure of a round object in n dimensions?

Some 50 years after my first exposure to the formulas for area and volume, I have finally had occasion to look into these broader questions. Finding the master formula for n-dimensional volumes was easy; a few minutes with Google and Wikipedia was all it took. But I've had many a brow-furrowing moment since then trying to make sense of what the formula is telling me. The relation between volume and dimension is not at all what I expected; indeed, it's one of the zaniest things I've ever come upon in mathematics. I'm appalled to realize that I have passed so much of my life in ignorance of this curious phenomenon. I write about it here in case anyone else also missed school on the day the class learned n-dimensional geometry.

Lost in Space

In those childhood years when I was memorizing volume formulas, I also played a lot of ball games. Often the game was delayed when we lost the ball in the weeds beyond right field. I didn't know it then, but we were lucky we played on a two-dimensional field. If we had lost our ball in a space of many dimensions, we might still be looking for it.

The mathematician Richard Bellman labeled this effect "the curse of dimensionality." As the number of spatial dimensions goes up, finding things or measuring their size and shape gets harder. This issue is a matter of practical consequence because many computational tasks are carried out in a high-dimensional setting. Typically each variable in a problem description is mapped to a separate dimension.

A few months ago I was preparing an illustration of Bellman's curse for an earlier Computing Science column. My first thought was to show the ball-in-a-box phenomenon. Put an n-dimensional ball in an n-dimensional cube just large enough to receive it. As n increases, the fraction of the cube's volume occupied by the ball falls dramatically.

In the end I chose a different and simpler scheme for the illustration. But after the column appeared ["Quasi-random Ramblings," *American Scientist*, July–August 2011], I returned to the ball-in-a-box question out of curiosity. I had long thought that I understood it, but I realized that I had almost no quantitative data on the relative size of the ball and the cube.

(In this context "ball" is not just a plaything but also the mathematical term for a solid spherical object. "Sphere" itself is generally reserved for a hollow shell, like a soap bubble. More formally, a sphere is the locus of all points whose distance from the center is equal to the radius r. A ball is the locus of points whose distance from the center is less than or equal to r. And while I'm trudging through this mire of terminology, I should mention that "n-ball" and "n-cube" refer to an n-dimensional object inhabiting n-dimensional space. This may seem too obvious to bother stating, but some branches of mathematics adopt a different convention. In topology, for instance, a 2-sphere lives in 3-space.)

The Master Formula

An n-ball of radius 1 (a "unit ball") just fits inside an n-cube with sides of length 2. The surface of the ball kisses the center of each face of the cube. In this configuration, what fraction of the cubic volume is filled by the ball?

The question is answered easily in the familiar low-dimensional spaces that we are all accustomed to living in. At the bottom of the hierarchy is one-dimensional geometry, which is rather dull: Everything looks like a line segment. A 1-ball with $r = 1$ and a 1-cube with $s = 2$ are actually

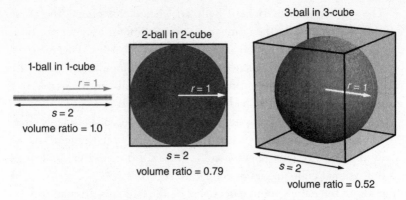

Figure 1. Balls in boxes offer a simple system for studying geometry across a series of spatial dimensions. A ball is the solid object bounded by a sphere; the boxes are cubes with sides of length 2, which makes them just large enough to accommodate a ball of radius 1. In one dimension (left) the ball and the cube have the same shape: a line segment of length 2. In two dimensions (center) and three dimensions (right) the ball and cube are more recognizable. As dimension increases, the ball fills a smaller and smaller fraction of the cube's internal volume. In three dimensions the filled fraction is about half; in 100-dimensional space, the ball has all but vanished, filling only 1.8×10^{-70} of the cube's volume.

the same object—a line segment of length 2. Thus in one dimension the ball completely fills the cube; the volume ratio is 1.0.

In two dimensions, a 2-ball inside a 2-cube is a disk inscribed in a square, and so this problem can be solved with one of my childhood formulas. With $r = 1$, the area πr^2 is simply π, whereas the area of the square, s^2, is 4; the ratio of these quantities is about 0.79.

In three dimensions, the ball's volume is $\frac{4}{3}\pi$, whereas the cube has a volume of 8; this works out to a ratio of approximately 0.52.

On the basis of these three data points, it appears that the ball fills a smaller and smaller fraction of the cube as n increases. There's a simple, intuitive argument suggesting that the trend will continue. The regions of the cube that are left vacant by the ball are the corners. Each time n increases by 1, the number of corners doubles, so we can expect ever more volume to migrate into the nooks and crannies near the cube's vertices.

To go beyond this appealing but nonquantitative principle, I would have to calculate the volume of n-balls and n-cubes for values of n

greater than 3. The calculation is easy for the cube. An n-cube with sides of length s has volume s^n. The cube that encloses a unit ball has $s = 2$, so the volume is 2^n.

But what about the n-ball? As I have already noted, my early education failed to equip me with the necessary formula, and so I turned to the Web. What a marvel it is! (And it gets better all the time.) In two or three clicks I had before me a Wikipedia page titled "Deriving the volume of an n-ball." Near the top of that page was the formula I sought:

$$V(n,r) = \frac{\pi^{\frac{n}{2}} r^n}{\Gamma(\frac{n}{2}+1)}$$

Later in this column I'll say a few words about where this formula came from, both mathematically and historically, but for now I merely note that the only part of the formula that ventures beyond routine arithmetic is the gamma function, Γ, which is an elaboration on the idea of a factorial. For positive integers, $\Gamma(n+1) = n! = 1 \times 2 \times 3 \times \ldots \times n$. But the gamma function, unlike the factorial, is also defined for numbers other than integers. For example, $\Gamma(\frac{1}{2})$ is equal to $\sqrt{\pi}$.

The Incredible Shrinking n-Ball

When I discovered the n-ball formula, I did not pause to investigate its provenance or derivation. I was impatient to plug in some numbers and see what would come out. So I wrote a hasty one-line program in Mathematica and began tabulating the volume of a unit ball in various dimensions. I had definite expectations about the outcome. I believed that the volume of the unit ball would increase steadily with n, though at a lower rate than the volume of the enclosing $s = 2$ cube, thereby confirming Bellman's curse of dimensionality. Here are the first few results returned by the program:

n	$V(n,1)$
1	2
2	$\pi \approx 3.1416$
3	$\frac{4}{3}\pi \approx 4.1888$
4	$\frac{1}{2}\pi^2 \approx 4.9348$
5	$\frac{8}{15}\pi^2 \approx 5.2638$

I noted immediately that the values for one, two, and three dimensions agreed with the results I already knew. (This kind of confirmation is always reassuring when you run a program for the first time.) I also observed that the volume was slowly increasing with n, as I had expected.

But then I looked at the continuation of the table:

n	$V(n,1)$
1	2
2	$\pi \approx 3.1416$
3	$\frac{4}{3}\pi \approx 4.1888$
4	$\frac{1}{2}\pi^2 \approx 4.9348$
5	$\frac{8}{15}\pi^2 \approx 5.2638$
6	$\frac{1}{6}\pi^3 \approx 5.1677$
7	$\frac{16}{105}\pi^3 \approx 4.7248$
8	$\frac{1}{24}\pi^4 \approx 4.0587$
9	$\frac{32}{945}\pi^4 \approx 3.2985$
10	$\frac{1}{120}\pi^5 \approx 2.5502$

Beyond the fifth dimension, the volume of a unit n-ball *decreases* as n increases! I tried a few larger values of n, finding that $V(20, 1)$ is about 0.0258, and $V(100, 1)$ is in the neighborhood of 10^{-40}. Thus it looked very much like the n-ball dwindles away to nothing as n approaches infinity.

Doubly Cursed

I had thought that I understood Bellman's curse: Both the n-ball and the n-cube grow along with n, but the cube expands faster. In fact, the curse is far more damning: At the same time the cube inflates exponentially, the ball shrinks to insignificance. In a space of 100 dimensions, the fraction of the cubic volume filled by the ball has declined to 1.8×10^{-70}. This fraction is far smaller than the volume of an atom in relation to the volume of the Earth. The ball in the box has all but vanished. If you were to select a trillion points at random from the interior of the cube, you'd have almost no chance of landing on even one point that is also inside the ball.

What makes this disappearing act so extraordinary is that the ball in question is still the largest one that could possibly be stuffed into the cube. We are not talking about a pea rattling around loose inside a refrigerator carton. The ball's diameter is still equal to the side length of the cube. The surface of the ball touches every face of the cube. (A face of an n-cube is an $(n-1)$-cube.) The fit is snug; if the ball were made even a smidgen larger, it would bulge out of the cube on all sides. Nevertheless, in terms of volume measure, the ball is nearly crushed out of existence, like a black hole collapsing under its own mass.

How can we make sense of this seeming paradox? One way of understanding it is to acknowledge that the ball fills the middle of the cube, but the cube doesn't have much of a middle; almost all of its volume is away from the center, huddling in the corners. A simple counting argument gives a clue to what's going on. As noted above, the ball touches the enclosing cube at the center of each face, but it does not reach out into the corners. A 100-cube has just 200 faces, but it has 2^{100} corners.

Another approach to understanding the collapse of the n-ball is to imagine poking skewers through the cube along various diameters. (A diameter is any straight line that passes through the center point.) The shortest diameters run from the center of a face to the center of the opposite face. For the cube enclosing a unit ball, the length of this shortest diameter is 2, which is both the side length of the cube and the diameter of the ball. Thus a skewer on the shortest diameter lies inside the ball throughout its length.

The longest diameters of the cube extend from a corner through the center point to the opposite corner. For an n-cube with side length $s = 2$, the length of this diameter is $2\sqrt{n}$. Thus in the 100-cube surrounding a unit ball, the longest diameter has length 20; only 10 percent of this length lies within the ball. Moreover, there are just 100 of the shortest diameters, but there are 2^{99} of the longest ones.

Here is still another mind-bending trick with balls and boxes to suggest just how weird space becomes in higher dimensions. I learned of it from Barry Cipra, who published a description in Volume 1 of *What's Happening in the Mathematical Sciences* (1991). On the plane, a square with sides of length 4 accommodates four unit disks in a two-by-two array, with room for a smaller disk in the middle; the radius of that smaller disk is $\sqrt{2} - 1$. In three dimensions, the equivalent 3-cube fits eight unit balls, plus a smaller ninth ball in the middle, whose radius is $\sqrt{3} - 1$. In the general case of n dimensions, the box has room for 2^n unit n-balls in a rectilinear array, with one additional ball in the vacant central space, and the central ball has a radius of $\sqrt{n} - 1$. Look what happens when n reaches 9. The "smaller" central ball now has a radius of 2, which makes it twice the size of the 512 surrounding balls.

Furthermore, the central ball has expanded to reach the sides of the
bounding box, and it will burst through the walls with any further
increase in dimension.

What's So Special About the 5-Ball?

I was taken by surprise when I learned that the volume of a unit n-
ball goes to zero as n goes to infinity; I had expected the opposite.
But something else surprised me even more—the fact that the volume
function is not monotonic. Either a steady increase or a steady decrease
seemed more plausible than having the volume grow for a while, then
reach a peak at some finite value of n, and thereafter decline. This behavior singles out a particular dimension for special attention. What is
it about five-dimensional space that allows a unit 5-ball to spread out
more expansively than any other n-ball?

I can offer an answer, although it doesn't really explain much. The
answer is that everything depends on the value of π. Because π is a little
more than 3, the volume peak comes in five dimensions; if π were equal

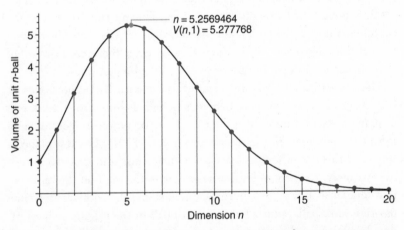

Figure 2. The volume of a unit ball in n dimensions reveals an intriguing spectrum of variations. Up to dimension 5, the ball's volume increases with each
increment to n; then the volume starts diminishing again, and ultimately goes
to zero as n goes to infinity. If dimension is considered a continuous variable,
the peak volume comes at $n = 5.2569464$ (the dot at the top of the curve).

to 17, say, the unit ball with maximum volume would be found in a space with 33 dimensions.

To see how π comes to have this role, we'll have to return to the formula for n-ball volume. We can get a rough sense of the function's behavior from a simplified version of the formula. In the first place, if we are interested only in the unit ball, then r is always equal to 1, and the r^n term can be ignored. That leaves a power of π in the numerator and a gamma function in the denominator. If we consider only even values of n, so that $n/2$ is always an integer, we can replace the gamma function with a factorial. For brevity, let $m = n/2$; then all that remains of the formula is this ratio: $\pi^m/m!$.

The simplified formula says that the n-ball volume is determined by a race between π^m and $m!$. Initially, for the smallest values of m, π^m sprints ahead; for example, at $m = 2$ we have $\pi^2 \approx 10$, which is greater than $2! = 2$. In the long run, however, $m!$ will surely win this race. Both π^m and $m!$ are products of m factors, but in π^m the factors are all equal to π, whereas in $m!$ they range from 1 up to m. Numerically, $m!$ first exceeds π^m when $m = 7$, and thereafter the factorial grows much larger.

This simplified analysis accounts for the major features of the volume curve, at least in a qualitative way. The volume of a unit ball has to go to zero in infinite-dimensional space because zero is the limit of the ratio $\pi^m/m!$. In low dimensions, on the other hand, the ratio is increasing with m. And if it's going uphill for small m and downhill for large m, there must be some intermediate value where the function reaches a maximum.

To get a quantitative fix on the location of the maximum, we must return to the formula in its original form and consider odd as well as even numbers of dimensions. Indeed, we can take a step beyond mere integer dimensions. Because the gamma function is defined for all real numbers, we can treat dimension as a continuous variable and ask with finer resolution where the maximum volume occurs. A numerical solution to this calculus problem—found with further help from Mathematica—shows a peak in the volume curve at $n \approx 5.2569464$; at this point the unit ball has a volume of 5.2777680.

With a closely related formula, we can also calculate the surface area of an n-ball. Like the volume, this quantity reaches a peak and then falls away to zero. The maximum is at $n \approx 7.2569464$, or in other words two dimensions larger than the volume peak.

The Dimensions of the Problem

The arithmetic behind all these results is straightforward; attaching meaning to the numbers is not so easy. In particular, I can see numerically—by comparing powers of π with factorials—why the unit ball's volume reaches a maximum at $n = 5$. But I have no geometric intuition about five-dimensional space that would explain this fact. Perhaps readers with deeper vision may be able to provide some insight.

The results on noninteger dimensions are quite otherworldly. The notion of fractional dimensions is familiar enough, but it is generally applied to objects, not to spaces. For example, the Sierpinski triangle, with its endlessly nested holes within holes, is assigned a dimension of 1.585, but the triangle is still drawn on a plane of dimension 2. What would it mean to construct a space with 5.2569464 mutually perpendicular coordinate axes? I can't imagine—and that's not just a figure of speech.

Another troubling question is whether it really makes sense to compare volumes across dimensions. Each dimension requires its own units of measure, and so the relative magnitudes of the numbers attached to those units don't mean much. Is a disk of area 10 cm^2 larger or smaller than a ball of volume 5 cm^3? We can't answer; it's like comparing apples and orange juice.

Nevertheless, I believe that there is indeed a valid basis for making comparisons. In each dimension, volume is to be measured in terms of a standard volume *in that dimension*. The obvious standard is the unit cube (sometimes called the "measure polytope"), which has a volume of 1 in all dimensions. Starting at $n = 1$, the unit ball is larger than the unit cube, and the ball-to-cube ratio gets still larger through $n = 5$; then the trend reverses, and eventually the ball is much smaller than the unit cube. This changing ratio of ball volume to cube volume is the phenomenon to be explained.

Slicing the Onion

The volume formulas I learned as a child were incantations to be memorized rather than understood. I would like to do better now. Although I cannot give a full derivation of the n-ball formula—for lack of both space and mathematical acumen—perhaps the following remarks may shed some light.

The key idea is that an n-ball has within it an infinity of $(n-1)$-balls. For example, a series of parallel slices through the body of an onion turns a 3-ball into a stack of 2-balls. Another set of cuts, perpendicular to the first series, reduces each disklike slice to a collection of 1-balls— linear ribbons of onion. If you go on to dice the ribbons, you have a heap of 0-balls. (With real onions and knives, these operations only approximate the forms of true n-balls, but the methods work perfectly in the mathematical kitchen.)

This decomposition suggests a recursive algorithm for computing the volume of an n-ball: Slice it into many $(n-1)$-balls and sum up the volumes of the slices. How do you compute the volumes of the slices? Apply the same method, cutting the $(n-1)$-balls into $(n-2)$-balls. Eventually the recursion bottoms out at $n = 1$ or $n = 0$, where the answers are known. (The volume of a 1-ball is $2r$; the 0-ball is assigned a volume of 1.) Letting the thickness of the slices go to zero turns the sum into an integral and leads to an exact result.

In practice, it's convenient to use a slightly different recursion with a step size of 2. That is, the volume of an n-ball is computed from that of an $(n-2)$-ball. The specific rule is this: Given the volume of an $(n-2)$-ball, multiply by $2\pi r^2/n$ to get the volume of the corresponding n-ball. (Showing *why* the multiplicative factor takes this particular form is the hard part of the derivation, which I am going to gingerly avoid; it requires an exercise in multivariable calculus that lies beyond my abilities.)

The procedure is easy to express in the form of a computer program:

function $V(n,r)$
   if $n = 0$ then return 1
   else if $n = 1$ then return $2r$
   else return
     $2\pi r^2/n \times V(n-2,r)$

For even n, the sequence of operations carried out by this program amounts to

$$1 \times \frac{2\pi r^2}{2} \times \frac{2\pi r^2}{4} \times \frac{2\pi r^2}{6} \times \cdots \times \frac{2\pi r^2}{n}.$$

For odd n, the result is instead the product of these terms:

$$2r \times \frac{2\pi r^2}{3} \times \frac{2\pi r^2}{5} \times \frac{2\pi r^2}{7} \times \cdots \times \frac{2\pi r^2}{n}.$$

For all integer values of *n*, the program yields the same output as the formula based on the gamma function.

Who Done It?

A question I cannot answer with certainty is who first wrote down the *n*-ball formula. I have paddled up a long river of references, but I'm not sure I have reached the true source.

My journey began with the number 5.2569464. I entered the digits into the On-Line Encyclopedia of Integer Sequences, the vast compendium of number lore created by Neil J. A. Sloane. I found what I was looking for in sequence A074455. A reference there directed me to *Sphere Packings, Lattices, and Groups*, by John Horton Conway and Sloane. That book in turn cited *An Introduction to the Geometry of N Dimensions*, by Duncan Sommerville, published in 1929. The Sommerville book devotes a few pages to the *n*-ball formula and has a table of values for dimensions 1 through 7, but it says little about origins. However, further rooting in library catalogs revealed that Sommerville—a Scottish mathematician who emigrated to New Zealand in 1915—also published a bibliography of non-Euclidean and *n*-dimensional geometry.

The bibliography lists five works on "hypersphere volume and surface"; the earliest is a problem and solution published in 1866 by William Kingdon Clifford, a brilliant English geometer who died young. Clifford's derivation of the formula is clearly original work, but it was not the first.

Elsewhere Sommerville mentions the Swiss mathematician Ludwig Schläfli as a pioneer of *n*-dimensional geometry. Schläfli's treatise on the subject, written in the early 1850s, was not published in full until 1901, but an excerpt translated into English by Arthur Cayley appeared in 1858. The first paragraph of that excerpt gives the volume formula for an *n*-ball, commenting that it was determined "long ago." An asterisk leads to a footnote citing papers published in 1839 and 1841 by the Belgian mathematician Eugène Catalan.

Looking up Catalan's articles, I found that neither of them gives the correct formula in full, although they're close. Catalan deserves partial credit.

Not one of these early works pauses to comment on the implications of the formula—the peak at *n* = 5 or the trend toward zero volume in

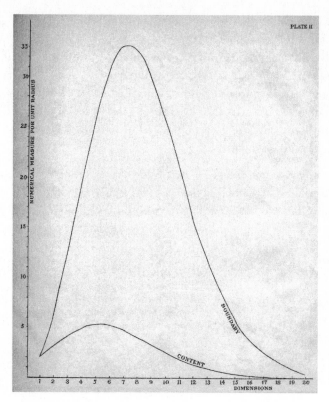

Figure 3. The graph of n-ball volume as a function of dimension was plotted more than 100 years ago by Paul Renno Heyl, who was then a graduate student at the University of Pennsylvania. The volume graph is the lower curve, labeled "content." The upper curve gives the ball's surface area, for which Heyl used the term "boundary." The illustration is from Heyl's 1897 thesis, "Properties of the locus *r* = constant in space of n dimensions."

high dimensions. Of the works mentioned by Sommerville, the only one to make these connections is a thesis by Paul Renno Heyl, published by the University of Pennsylvania in 1897. This looked like a fairly obscure item, but with help from Harvard librarians, the volume was found on a basement shelf. I later discovered that the full text (but not the plates) is available on Google Books.

Heyl was a graduate student at the time of this work. He went on to a career with the National Bureau of Standards, and he was also a writer on science, philosophy, and religion. (His best-known book was _The Mystery of Evil_.)

In the 1897 thesis, Heyl derives formulas for both volume and surface area (which he calls "content" and "boundary") and gives a lucid account of multidimensional geometry in general. He clearly appreciates the strangeness of the discovery that ". . . in a space of infinite dimension our locus can have no content at all." I will allow Heyl to have the last word on the subject:

> We might be pardoned for supposing that in a space of infinite dimension we should find the Absolute and Unconditioned if anywhere, but we have reached an opposite conclusion. This is the most curious thing I know of in the Wonderland of Higher Space.

Bibliography

Ball, K. 1997. An elementary introduction to modern convex geometry. In _Flavors of Geometry_. Silvio Levy (ed.), Cambridge: Cambridge University Press.

Bellman, R. E. 1961. _Adaptive Control Processes: A Guided Tour_. Princeton: Princeton University Press.

Catalan, Eugène. 1839, 1841. _Journal de Mathématiques Pures et Appliquées_ 4:323–344, 6:81–84.

Cipra, B. 1991. Here's looking at Euclid. In _What's Happening in the Mathematical Sciences_, Vol. 1, p. 25. Providence: American Mathematical Society.

Clifford, W. K. 1866. Question 1878. _Mathematical Questions, with Their Solutions, from the "Educational Times"_ 6:83–87.

Conway, J. H., and N.J.A. Sloane. 1999. _Sphere Packings, Lattices, and Groups_. 3rd edition. New York: Springer.

Heyl, P. R. 1897. Properties of the locus $r =$ constant in space of n dimensions. Philadelphia: Publications of the University of Pennsylvania, Mathematics, No. 1, pp. 33–39. Available online at http://books.google.com/books?id=j5pQAAAAYAAJ

On-Line Encyclopedia of Integer Sequences, published electronically at http://oeis.org, 2010, Sequence A074455.

Schläfli, L. 1858. On a multiple integral. _The Quarterly Journal of Pure and Applied Mathematics_ 2:269–301.

Sommerville, D.M.Y. 1911. _Bibliography of Non-Euclidean Geometry, Including the Theory of Parallels, the Foundation of Geometry, and Space of N Dimensions_. London: Harrison & Sons.

Sommerville, D.M.Y. 1929. _An Introduction to the Geometry of N Dimensions_. New York: Dover Publications.

Wikipedia. Deriving the volume of an n-ball. http://en.wikipedia.org/wiki/Deriving_the _volume_of_an_n-ball

Structure and Randomness in the Prime Numbers

TERENCE TAO

The prime numbers 2, 3, 5, 7, . . . are one of the oldest topics studied in mathematics. We now have a lot of intuition as to how the primes *should* behave, and a great deal of confidence in our conjectures about the primes . . . but we still have a great deal of difficulty in *proving* many of these conjectures! Ultimately, this difficulty occurs because the primes are believed to behave *pseudorandomly* in many ways, and not to follow any simple pattern. We have many ways of establishing that a pattern exists . . . but how does one demonstrate the absence of a pattern?

In this chapter, I try to convince you why the primes are believed to behave pseudorandomly and tell you how one could try to make this intuition rigorous. This is only a small sample of what is going on in the subject; I am omitting many major topics, such as sieve theory or exponential sums, and I am glossing over many important technical details.

Finding Primes

It is a paradoxical fact that the primes are simultaneously numerous and hard to find. First, we have the following ancient theorem from Euclid [2]:

There are infinitely many primes.

In particular, given any k, there exists a prime with at least k digits. But there is no known *quick* and *deterministic* way to locate such a prime! (Here, "quick" means "computable in a time which is polynomial in k".) In particular, there is no known (deterministic) formula that can quickly generate large numbers that are guaranteed to be prime. Currently, the largest known prime is $2^{43,112,609} - 1$, about 13 million digits long [3].

On the other hand, one can find primes quickly by *probabilistic* methods. Indeed, any k-digit number can be tested for primality quickly, either by probabilistic methods [10, 12] or by deterministic methods [1]. These methods are based on variants of Fermat's little theorem, which asserts that $a^n \equiv a \bmod n$ whenever n is prime. (Note that $a^n \bmod n$ can be computed quickly, by first repeatedly squaring a to compute $a^{2^j} \bmod n$ for various values of j and then expanding n in binary and multiplying the indicated residues $a^{2^j} \bmod n$ together.)

Also, we have the following fundamental theorem of prime numbers [8, 14, 16]:

> *The number of primes less than a given integer n is $(1 + o(1))\frac{n}{\log n}$, where $o(1)$ tends to zero as $n \to \infty$.*

(We use log to denote the natural logarithm.) In particular, the probability of a randomly selected k-digit number being prime is about $\frac{1}{k \log 10}$. So one can quickly find a k-digit prime with high probability by randomly selecting k-digit numbers and testing each of them for primality.

Is Randomness Really Necessary?

To summarize: We do not know a quick way to find primes *deterministically*. However, we have quick ways to find primes *randomly*.

On the other hand, there are major conjectures in complexity theory, such as P = BPP, which assert (roughly speaking) that any problem that can be solved quickly by probabilistic methods can also be solved quickly by deterministic methods.[1]

These conjectures are closely related to the more famous conjecture P $\neq$ NP, which is a US\$1 million Clay Millennium prize problem.[2]

Many other important probabilistic algorithms have been *derandomized* into deterministic ones, but this *derandomization* has not been done for the problem of finding primes. (A massively collaborative research project is currently underway to attempt this solution [11].)

Counting Primes

We've seen that it's hard to get a hold of any single large prime. But it is easier to study the set of primes *collectively* rather than one at a time.

An analogy: it is difficult to locate and count all the grains of sand in a box, but one can get an estimate on this count by *weighing* the box, subtracting the weight of the empty box, and dividing by the average weight of a grain of sand. The point is that there is an easily measured statistic (the weight of the box with the sand) that reflects the *collective* behavior of the sand.

For instance, from the *fundamental theorem of arithmetic* one can establish Euler's *product formula*

$$\sum_{n=1}^{\infty} \frac{1}{n^s} = \prod_{p \text{ prime}} \left(1 + \frac{1}{p^s} + \frac{1}{p^{2s}} + \frac{1}{p^{3s}} + \ldots\right) = \prod_{p \text{ prime}} \left(1 - \frac{1}{p^s}\right)^{-1} \quad (1)$$

for any $s > 1$ (and also for other complex values of s, if one defines one's terms carefully enough).

The formula (1) links the collective behavior of the primes to the behavior of the *Riemann zeta function*

$$\zeta(s) := \sum_{n=1}^{\infty} \frac{1}{n^s},$$

thus

$$\prod_{p \text{ prime}} \left(1 - \frac{1}{p^s}\right) = \frac{1}{\zeta(s)}. \quad (2)$$

One can then deduce information about the primes from information about the zeta function (and in particular, its zeros).

For instance, from the divergence of the harmonic series $\sum_{n=1}^{\infty} \frac{1}{n} = +\infty$ we see that $\frac{1}{\zeta(s)}$ goes to zero as s approaches 1 (from the right, at least). From this and Equation (2), we already recover Euclid's theorem (Theorem 1), and in fact obtain the stronger result of Euler that the sum $\sum_p \frac{1}{p}$ of reciprocals of primes diverges also.[3]

In a similar spirit, one can use the techniques of complex analysis, combined with the (nontrivial) fact that $\zeta(s)$ is never zero for $s \in \mathbb{C}$ when $\text{Re}(s) \geq 1$, to establish the prime number theorem [16]; indeed, this is how the theorem was originally proved [8, 14] (and one can conversely use the prime number theorem to deduce the fact about the zeros of ζ).

The famous *Riemann hypothesis* asserts that $\zeta(s)$ is never zero when[4] $\text{Re}(s) > 1/2$. It implies a much stronger version of the prime number theorem, namely that the number of primes less than an integer $n > 1$

is given by the more precise formula[5] $\int_0^n \frac{dx}{\log x} + O(n^{1/2} \log n)$, where $O(n^{1/2} \log n)$ is a quantity that is bounded in magnitude by $Cn^{1/2} \log n$ for some absolute constant C (for instance, one can take $C = \frac{1}{8\pi}$ once n is at least 2657 [13]). The hypothesis has many other consequences in number theory; it is another of the US$1 million Clay Millennium prize problems. More generally, much of what we know about the primes has come from an extensive study of the properties of the Riemann zeta function and its relatives, although there are also some questions about primes that remain out of reach even assuming strong conjectures such as the Riemann hypothesis.

Modeling Primes

A fruitful way to think about the set of primes is as a *pseudorandom set*—a set of numbers that is not actually random but behaves like a random set.

For instance, the prime number theorem asserts, roughly speaking, that a randomly chosen large integer n has a probability of about $1/(\log n)$ of being prime. One can then *model* the set of primes by replacing them with a random set of integers in which each integer $n > 1$ is selected with an independent probability of $1/(\log n)$; this is *Cramér's random model*.

This model is too crude because it misses some obvious structure in the primes, such as the fact that most primes are odd. But one can improve the model to address this issue by picking a model where odd integers n are selected with an independent probability of $2/(\log n)$ and even integers are selected with probability 0.

One can also take into account other obvious structure in the primes, such as the fact that most primes are not divisible by 3, not divisible by 5, etc. This situation leads to fancier random models, which we believe to accurately predict the asymptotic behavior of primes.

For example, suppose we want to predict the number of twin primes n, $n + 2$, where $n \le N$ for a given threshold N. Using the Cramér random model, we expect, for any given n, that n, $n + 2$ will simultaneously be prime with probability $\frac{1}{\log n \log(n+2)}$, so we expect the number of twin primes to be about[6]

$$\sum_{n=1}^{N} \frac{1}{\log n \log(n + 2)} \approx \frac{N}{\log^2 N}.$$

This prediction is inaccurate; for instance, the same argument would also predict plenty of pairs of *consecutive* primes n, $n + 1$, which is absurd. But if one uses the refined model where odd integers n are prime with an independent probability of $2/\log n$ and even integers are prime with probability 0, one gets the slightly different prediction

$$\sum_{\substack{1 \le n \le N \\ n \text{ odd}}} \frac{2}{\log n} \times \frac{2}{\log(n+2)} \approx 2 \frac{N}{\log^2 N}.$$

More generally, if one assumes that all numbers n divisible by some prime less than a small threshold w are prime with probability zero, and are prime with a probability of $\prod_{p<w} (1-\frac{1}{p})^{-1} \times \frac{1}{\log n}$; otherwise, one is eventually led to the prediction

$$2\left(\prod_{\substack{p<w \\ p \text{ odd}}} \frac{p-2}{p}\left(1-\frac{1}{p}\right)^{-2}\right)\frac{N}{\log^2 N} = 2\left(\prod_{\substack{p<w \\ p \text{ odd}}}\left(1-\frac{1}{(p-1)^2}\right)\right)\frac{N}{\log^2 N}$$

(for p an odd prime, among p consecutive integers, only $p-2$ have a chance to be the smaller number in a pair of twin primes). Sending $w \to \infty$, one is led to the asymptotic prediction

$$\Pi_2 \frac{N}{\log^2 N}$$

for the number of twin primes less than N, where Π_2 is the *twin prime constant*

$$\Pi_2 := 2 \prod_{p \text{ odd prime}} \left(1-\frac{1}{(p-1)^2}\right) \approx 1.32032\ldots.$$

For $N = 10^{10}$, this prediction is accurate to four decimal places and is believed to be asymptotically correct. (This is part of a more general conjecture, known as the *Hardy–Littlewood prime tuples conjecture* [9].)

Similar arguments based on random models give convincing heuristic support for many other conjectures in number theory and are backed up by extensive numerical calculations.

Finding Patterns in Primes

Of course, the primes are a deterministic set of integers, not a random one, so the predictions given by random models are not rigorous. But can they be made so?

There has been some progress in doing this. One approach is to try to classify all the possible ways in which a set could fail to be pseudorandom (i.e., it does something noticeably different from what a random set would do) and then show that the primes do not behave in any of these ways.

For instance, consider the *odd Goldbach conjecture*: every odd integer larger than five is the sum of three primes. If, for instance, all large primes happened to have their last digit equal to one, then Goldbach's conjecture could well fail for some large odd integers whose last digit was different from three. Thus we see that the conjecture could fail if there was a sufficiently strange "conspiracy" among the primes.

However, one can rule out this particular conspiracy by using the *prime number theorem in arithmetic progressions*, which tells us that (among other things) there are many primes whose last digit is different from 1. (The proof of this theorem is based on the proof of the classical prime number theorem.)

Moreover, by using the techniques of *Fourier analysis* (or more precisely, the *Hardy–Littlewood circle method*), we can show that *all* the conspiracies which could conceivably sink Goldbach's conjecture (for large integers, at least) are broadly of this type: an unexpected "bias" for the primes to prefer one remainder modulo 10 (or modulo another base, which need not be an integer) over another.

Vinogradov [15] eliminated each of these potential conspiracies and established *Vinogradov's theorem*: Every sufficiently large odd integer is the sum of three primes.[7] This method has since been extended by many authors to cover many other types of patterns; for instance, related techniques were used by Ben Green and myself [4] to establish that the primes contain arbitrarily long arithmetic progressions, and in subsequent work of Ben Green, myself, and Tamar Ziegler [5, 6, 7] to count a wide range of other additive patterns also. (Very roughly speaking, known techniques can count additive patterns that involve two independent parameters, such as arithmetic progressions a, $a + r$, . . . , $a + (k − 1)r$ of a fixed length k.)

Unfortunately, "one-parameter" patterns, such as twins n, $n + 2$, remain stubbornly beyond current technology. There is still much to be done in the subject!

Notes

1. Strictly speaking, the P = BPP conjecture only applies to decision problems—problems with a yes/no answer—rather than search problems, such as the task of finding a prime, but there are variants of P = BPP, such as P = promise-BPP, which would be applicable here.

2. The precise definitions of P, NP, and BPP are quite technical; suffice it to say that P stands for "polynomial time," NP stands for "nondeterministic polynomial time," and BPP stands for "bounded-error probabilistic polynomial time."

3. Observe that $\log(1/\zeta(s)) = \log \prod_p (1 - p^{-s}) = \sum_p \log(1 - p^{-s}) \geq -2\sum_p p^{-s}$.

4. A technical point: The sum $\sum_{n=1}^{\infty} \frac{1}{n^s}$ does not converge in the classical sense when $\mathrm{Re}(s) \leq 1$, so one has to interpret this sum in a fancier way, or else use a different definition of $\zeta(s)$ in this case; but I will not discuss these subtleties here.

5. The Prime Number Theorem says that, as $n \to \infty$, the number of correct decimal digits in the estimate $n/(\log n)$ tends to infinity, but it does not relate the number of correct digits to the total number of digits of $\pi(n)$. If the Riemann hypothesis is correct, then $\int_0^n dx/\log x$ correctly predicts almost half of the digits in $\pi(n)$.

6. We use the symbol $\approx$ in the sense that the quotient of the two quantities tends to 1 as $N \to \infty$.

7. Vinogradov himself could not specify explicitly what "sufficiently large" is. Soon after, his student Borozdin showed that numbers greater than $e^{3^{15}} \approx 10^{6\,846\,169}$ are "sufficiently large." Meanwhile, this bound has been lowered to $e^{3^{100}} \approx 10^{1\,346}$—still far beyond reach for computer tests for the smaller numbers.

References

[1] Manindra Agrawal, Neeraj Kayal, and Nitin Saxena, PRIMES is in P. *Annals of Mathematics* **160**, 781–793 (2004).

[2] Euclid, *The Elements*, circa 300 BCE.

[3] Great Internet Mersenne Prime Search. http://www.mersenne.org (2008).

[4] Ben Green and Terence Tao, The primes contain arbitrarily long arithmetic progressions. *Annals of Mathematics* **167**(2), 481–547 (2008).

[5] Ben Green and Terence Tao, *Linear equations in primes*. Preprint. http://arxiv.org/abs/math/0606088, 84 pages (April 22, 2008).

[6] Ben Green and Terence Tao, *The Möbius function is asymptotically orthogonal to nilsequences*. Preprint. http://arxiv.org/ahs/0807.1736, 22 pages (April 26, 2010).

[7] Ben Green, Terence Tao, and Tamar Ziegler, *The inverse conjecture for the Gowers norm*. Preprint.

[8] Jacques Hadamard, Sur la distribution des zéros de la fonction $\zeta(s)$ et ses conséquences arithmétiques. *Bulletin de la Société Mathématique de France* **24**, 199–220 (1896).

[9] Godfrey H. Hardy and John E. Littlewood, Some problems of "partitio numerorum." III. On the expression of a number as a sum of primes. *Acta Mathematica* **44**, 1–70 (1923).

[10] Gary L. Miller, Riemann's hypothesis and tests for primality. *Journal of Computer and System Sciences* **13**(3), 300–317 (1976).

[11] Polymath4 project: Deterministic way to find primes. http://michaelnielsen.org/polymath1/index.php?title=Finding_primes

[12] Michael O. Rabin, Probabilistic algorithm for testing primality. *Journal of Number Theory* **12**, 128–138 (1980).

[13] Lowell Schoenfeld, Sharper bounds for the Chebyshev functions $\theta(x)$ and $\psi(x)$. II. *Mathematics of Computation* **30**, 337–360 (1976).

[14] Charles-Jean de la Vallée Poussin, Recherches analytiques de la théorie des nombres premiers. *Annales de la Société scientifique de Bruxelles* **20**,183–256 (1896).

[15] Ivan M. Vinogradov, The method of trigonometrical sums in the theory of numbers (Russian). *Travaux de l'Institut Mathématique Stekloff* **10** (1937).

[16] Don Zagier, Newman's short proof of the prime number theorem. *American Mathematical Monthly* **104**(8), 705–708 (1997).

The Strangest Numbers in String Theory

JOHN C. BAEZ AND JOHN HUERTA

As children, we all learn about numbers. We start with counting, followed by addition, subtraction, multiplication, and division. But mathematicians know that the number system we study in school is but one of many possibilities. Other kinds of numbers are important for understanding geometry and physics. One of the strangest alternatives is the octonions. Largely neglected since their discovery in 1843, in the past few decades they have assumed a curious importance in string theory. And indeed, if string theory is a correct representation of the universe, they may be part of the reason the universe has the number of dimensions it does.

The Imaginary Made Real

The octonions would not be the first piece of pure mathematics that was later used to enhance our understanding of the cosmos. Nor would it be the first alternative number system that was later shown to have practical uses. To understand why, we first have to look at the simplest case of numbers—the number system we learned about in school—which mathematicians call the real numbers. The set of all real numbers forms a line, so we say that the collection of real numbers is one-dimensional. We could also turn this idea on its head: the line is one-dimensional because specifying a point on it requires one real number.

Before the 1500s the real numbers were the only game in town. Then, during the Renaissance, ambitious mathematicians attempted to solve ever more complex forms of equations, even holding competitions to see who could solve the most difficult problems. The square root of -1 was introduced as a kind of secret weapon by Italian mathematician, physician, gambler, and astrologer Girolamo Cardano. Where others might cavil, he boldly let himself use this mysterious number as part of longer calculations where the answers were ordinary real numbers. He was not sure why this trick worked; all he knew was that it gave him the right answers. He published his ideas in 1545, thus beginning a controversy that lasted for centuries: Does the square root of −1 really exist, or is it only a trick? Almost 100 years later, no less a thinker than René Descartes rendered his verdict when he gave it the derogatory name "imaginary," now abbreviated as i.

Nevertheless, mathematicians followed in Cardano's footsteps and began working with complex numbers—numbers of the form $a + bi$, where a and b are ordinary real numbers. Around 1806 Jean-Robert Argand popularized the idea that complex numbers describe points on the plane. How does $a + bi$ describe a point on the plane? Simple: The number a tells us how far left or right the point is, whereas b tells us how far up or down it is.

In this way, we can think of any complex number as a point in the plane, but Argand went a step further: He showed how to think of the operations one can do with complex numbers—addition, subtraction, multiplication, and division—as geometric manipulations in the plane (Figure 1).

As a warm-up for understanding how these operations can be thought of as geometric manipulations, first think about the real numbers. Adding or subtracting any real number slides the real line to the right or left. Multiplying or dividing by any positive number stretches or squashes the line. For example, multiplying by 2 stretches the line by a factor of 2, whereas dividing by 2 squashes it down, moving all the points twice as close as they were. Multiplying by −1 flips the line over.

The same procedure works for complex numbers, with just a few extra twists. Adding any complex number $a + bi$ to a point in the plane slides that point right (or left) by an amount a and up (or down) by an amount b. Multiplying by a complex number stretches or squashes but also rotates the complex plane. In particular, multiplying by i rotates the plane a quarter turn. Thus, if we multiply 1 by i twice, we rotate the plane a full half-turn from the starting point to arrive at −1. Division is the opposite of multiplication, so to divide we just shrink instead of stretching, or vice versa, and then rotate in the opposite direction.

Almost everything we can do with real numbers can also be done with complex numbers. In fact, most things work better, as Cardano

a. Real Numbers

Addition	**Subtraction**	**Multiplication**	**Division**
$0 + 2 = 2$	$0 - 2 = -2$	$2 \times 2 = 4$	$2 \div 2 = 1$

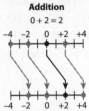

Addition along the real number line is simple: just shift each number to the right by the amount you are adding.

Subtraction operates the same way, but here we shift numbers to the left.

In multiplication, we stretch the number line out by a constant factor.

Division is equivalent to shrinking the points on the number line.

b. Complex Numbers

Addition	**Subtraction**	**Multiplication**	**Division**
$i + (2 + i) = 2 + 2i$	$i - (2 + i) = -2 + 0i$	$i \times (2i) = -2$	$2i \div (2i) = 1$

   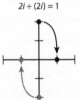

Complex numbers have two components—the real part, which is measured on the horizontal axis, and the imaginary part (noted by the *i*), which goes up the vertical axis. Adding two complex numbers shifts the original number to the *right* by the amount in the real part and *up* by the amount in the imaginary part.

Similarly, when we subtract complex numbers we shift the original point to the *left* by the amount in the real part and *down* by the amount in the imaginary part.

Multiplication is where the fun begins: Just as in the case of the real numbers, multiplication stretches a complex number. Moreover, multiplication by *i* rotates the point counterclockwise by 90 degrees.

Division shrinks a complex number, just as in the case of the real numbers. Division by *i* also rotates a complex number clockwise by 90 degrees.

Figure 1. Beyond the Real Line: Math in Multiple Dimensions. In grade school we are taught to connect the abstract ideas of addition and subtraction to concrete operations—moving numbers up and down the number line. This connection between algebra and geometry turns out to be incredibly powerful. Because of it, mathematicians can use the algebra of the octonions to solve problems in hard-to-imagine eight-dimensional worlds. The panels below show how to extend algebraic operations on the real-number line to complex (two-dimensional) numbers.

knew, because we can solve more equations with complex numbers than with real numbers. But if a two-dimensional number system gives the user added calculating power, what about even higher-dimensional systems? Unfortunately, a simple extension turns out to be impossible. An Irish mathematician would uncover the secret to higher-dimensional number systems decades later. And only now, two centuries on, are we beginning to understand how powerful they can be.

Hamilton's Alchemy

In 1835, at the age of 30, mathematician and physicist William Rowan Hamilton discovered how to treat complex numbers as pairs of real numbers. At the time, mathematicians commonly wrote complex numbers in the form $a + bi$ that Argand popularized, but Hamilton noted that we are also free to think of the number $a + bi$ as just a peculiar way of writing two real numbers—for instance (a, b).

This notation makes it easy to add and subtract complex numbers— just add or subtract the corresponding real numbers in the pair. Hamilton also came up with slightly more involved rules for how to multiply and divide complex numbers so that they maintained the nice geometric meaning discovered by Argand.

After Hamilton invented this algebraic system for complex numbers that had a geometric meaning, he tried for many years to invent a bigger algebra of triplets that would play a similar role in three-dimensional geometry, an effort that gave him no end of frustrations. He once wrote to his son, "Every morning . . . on my coming down to breakfast, your (then) little brother William Edwin, and yourself, used to ask me: 'Well, Papa, can you multiply triplets?' Whereto I was always obliged to reply, with a sad shake of the head: 'No, I can only add and subtract them.'" Although he could not have known it at the time, the task he had given himself was mathematically impossible.

Hamilton was searching for a three-dimensional number system in which he could add, subtract, multiply, and divide. Division is the hard part: A number system where we can divide is called a division algebra. Not until 1958 did three mathematicians prove an amazing fact that had been suspected for decades: Any division algebra must have dimension one (which is just the real numbers), two (the complex numbers), four, or eight. To succeed, Hamilton had to change the rules of the game.

Hamilton himself figured out a solution on October 16, 1843. He was walking with his wife along the Royal Canal to a meeting of the Royal Irish Academy in Dublin when he had a sudden revelation. In three dimensions, rotations, stretching, and shrinking could not be described with just three numbers. He needed a fourth number, thereby generating a four-dimensional set called quaternions that take the form $a + bi + cj + dk$. Here the numbers i, j, and k are three different square roots of -1.

Hamilton would later write, "I then and there felt the galvanic circuit of thought close; and the sparks which fell from it were the fundamental equations between i, j and k, exactly such as I have used them ever since." And in a noteworthy act of mathematical vandalism, he carved these equations into the stone of the Brougham Bridge. Although they are now buried under graffiti, a plaque has been placed there to commemorate the discovery.

It may seem odd that we need points in a four-dimensional space to describe changes in three-dimensional space, but it is true. Three of the numbers come from describing rotations, which we can see most readily if we imagine trying to fly an airplane. To orient the plane, we need to control the pitch, or angle with the horizontal. We also may need to adjust the yaw, by turning left or right, as a car does. And finally, we may need to adjust the roll: the angle of the plane's wings. The fourth number we need is used to describe stretching or shrinking.

Hamilton spent the rest of his life obsessed with the quaternions and found many practical uses for them. Today in many of these applications the quaternions have been replaced by their simpler cousins: vectors, which can be thought of as quaternions of the special form $ai + bj + ck$ (the first number is just zero). Yet quaternions still have their niche: They provide an efficient way to represent three-dimensional rotations on a computer and show up wherever these are needed, from the attitude-control system of a spacecraft to the graphics engine of a video game.

Imaginaries Without End

Despite these applications, we might wonder what, exactly, j and k are if we have already defined the square root of -1 as i. Do these square roots of -1 really exist? Can we just keep inventing new square roots of -1 to our heart's content?

These questions were asked by Hamilton's college friend, a lawyer named John Graves, whose amateur interest in algebra got Hamilton thinking about complex numbers and triplets in the first place. The very day after his fateful walk in the fall of 1843, Hamilton sent Graves a letter describing his breakthrough. Graves replied nine days later, complimenting Hamilton on the boldness of the idea but adding, "There is still something in the system which gravels me. I have not yet any clear views as to the extent to which we are at liberty arbitrarily to create imaginaries, and to endow them with supernatural properties." And he asked, "If with your alchemy you can make three pounds of gold, why should you stop there?"

Like Cardano before him, Graves set his concerns aside for long enough to conjure some gold of his own. On December 26 he wrote again to Hamilton, describing a new eight-dimensional number system that he called the octaves and that are now called octonions. Graves was unable to get Hamilton interested in his ideas, however. Hamilton promised to speak about Graves's octaves at the Irish Royal Society, which is one way mathematical results were published at the time. But Hamilton kept putting it off, and in 1845 the young genius Arthur Cayley rediscovered the octonions and beat Graves to publication. For this reason, the octonions are also sometimes known as Cayley numbers.

Why didn't Hamilton like the octonions? For one thing, he was obsessed with research on his own discovery, the quaternions. He also had a purely mathematical reason: The octonions break some cherished laws of arithmetic.

The quaternions were already a bit strange. When you multiply real numbers, it does not matter in which order you do it; 2 times 3 equals 3 times 2, for example. We say that multiplication commutes. The same holds for complex numbers. But quaternions are noncommutative. The order of multiplication matters.

Order is important because quaternions describe rotations in three dimensions, and for such rotations the order makes a difference to the outcome. You can check this out yourself (Figure 2). Take a book, flip it top to bottom (so that you are now viewing the back cover) and give it a quarter turn clockwise (as viewed from above). Now do these two operations in reverse order: first rotate a quarter turn, then flip. The final position has changed. Because the result depends on the order, rotations do not commute.

Flip, then rotate

Rotate, then flip

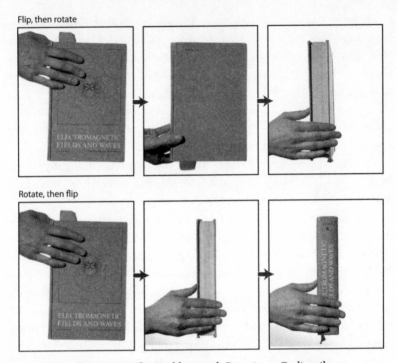

Figure 2. Visualizing 4-D: The Problem with Rotations. Ordinarily you can multiply numbers together in whatever order you like. For example, 2 times 3 is the same as 3 times 2. In higher dimensional number systems such as the quaternions and octonions, however, order is very important. Consider the quaternions, which describe rotations in three dimensions. If we take an object such as a book, the order in which we rotate it has a great effect on its final orientation. In the top row at the right, we flip the book vertically, then rotate it, revealing the page edges. In the bottom row, rotating the book and then flipping reveal the spine on the opposite side. Photo courtesy of Zachary Zavislak.

The octonions are much stranger. Not only are they noncommutative, they also break another familiar law of arithmetic: the associative law $(xy)z = x(yz)$. We have all seen a nonassociative operation in our study of mathematics: subtraction. For example, $(3 - 2) - 1$ is different from $3 - (2 - 1)$. But we are used to multiplication being associative, and most mathematicians still feel this way, even though they have gotten used to noncommutative operations. Rotations are associative, for example, even though they do not commute.

Perhaps most importantly, it was not clear in Hamilton's time just what the octonions would be good for. They are closely related to the geometry of seven and eight dimensions, and we can describe rotations in those dimensions using the multiplication of octonions. But for more than a century that was a purely intellectual exercise. It would take the development of modern particle physics—and string theory in particular—to see how the octonions might be useful in the real world.

Symmetry and Strings

In the 1970s and 1980s, theoretical physicists developed a strikingly beautiful idea called supersymmetry. (Later researchers would learn that string theory requires supersymmetry.) It states that at the most fundamental levels, the universe exhibits a symmetry between matter and the forces of nature. Every matter particle (such as an electron) has a partner particle that carries a force. And every force particle (such as a photon, the carrier of the electromagnetic force) has a twin matter particle.

Supersymmetry also encompasses the idea that the laws of physics would remain unchanged if we exchanged all the matter and force particles. Imagine viewing the universe in a strange mirror that, rather than interchanging left and right, traded every force particle for a matter particle, and vice versa. If supersymmetry is true, if it truly describes our universe, this mirror universe would act the same as ours. Even though physicists have not yet found any concrete experimental evidence in support of supersymmetry, the theory is so seductively beautiful and has led to so much enchanting mathematics that many physicists hope and expect that it is real.

One thing we know to be true, however, is quantum mechanics. And according to quantum mechanics, particles are also waves. In the standard three-dimensional version of quantum mechanics that physicists use every day, one type of number (called spinors) describes the wave motion of matter particles. Another type of number (called vectors) describes the wave motion of force particles. If we want to understand particle interactions, we have to combine these two using a curious way of multiplying a spinor and a vector to get a spinor. Although this system works, it does not treat matter and forces in a unified, elegant way.

As an alternative, imagine a strange universe with no time, only space. If this universe has dimension one, two, four, or eight, both matter and force particles would be waves described by a single type of number—namely, a number in a division algebra, the only type of system that allows for addition, subtraction, multiplication, and division. In other words, in these dimensions the vectors and spinors coincide: they are each just real numbers, complex numbers, quaternions, or octonions, respectively. Supersymmetry emerges naturally, providing a unified description of matter and forces. Simple multiplication describes interactions, and all particles—no matter the type—use the same number system.

Yet our plaything universe cannot be real, because we need to take time into account. In string theory, this consideration has an intriguing effect. At any moment in time, a string is a one-dimensional thing, like a curve or line. But this string traces out a two-dimensional surface as time passes (Figure 3). This evolution changes the dimensions in which supersymmetry arises, by adding two—one for the string and one for time. Instead of supersymmetry in dimension one, two, four, or eight, we get supersymmetry in dimension three, four, six, or ten.

Coincidentally, string theorists have for years been saying that only 10-dimensional versions of the theory are self-consistent. The rest suffer from glitches called anomalies, where computing the same thing in two different ways gives different answers. In anything other than 10 dimensions, string theory breaks down. But 10-dimensional string theory is, as we have just seen, the version of the theory that uses octonions. So if string theory is right, the octonions are not a useless curiosity. On the contrary, in 10 dimensions, matter and force particles are embodied in the same type of numbers—the octonions.

But this is not the end of the story. Recently physicists have started to go beyond strings to consider membranes. For example, a two-dimensional membrane, or 2-brane, looks like a sheet at any instant. As time passes, it traces out a three-dimensional volume in space-time.

Whereas in string theory we had to add two dimensions to our standard collection of one, two, four, and eight, now we must add three. Thus, when we are dealing with membranes, we would expect supersymmetry to naturally emerge in dimensions four, five, seven, and 11. And as in string theory, we have a surprise in store. Researchers tell us that M-theory (the "M" typically stands for "membrane") requires 11

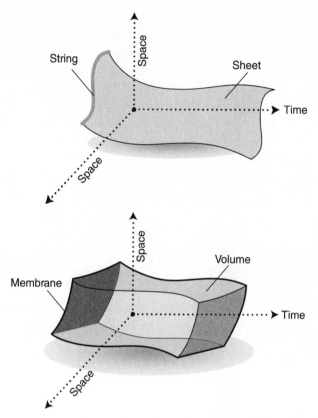

Figure 3. In string theory, one-dimensional strings trace out two-dimensional surfaces over time. In M-theory, two-dimensional membranes trace out three-dimensional volumes. Adding these dimensions to the eight dimensions of the octonions provides clues as to why these theories require 10 or 11 dimensions.

dimensions—implying that it should naturally make use of octonions. Alas, nobody understands M-theory well enough to even write down its basic equations (that M can also stand for "mysterious"). It is hard to tell precisely what shape it might take in the future.

At this point, we should emphasize that string theory and M-theory have not yet made any experimentally testable predictions. They are beautiful dreams—but so far only dreams. The universe we live in does not look 10- or 11-dimensional, and we have not seen any symmetry between matter and force particles. A while ago, David Gross, one of the world's leading experts on string theory, put the odds of seeing some evidence for supersymmetry at CERN's Large Hadron Collider at 50 percent. Skeptics say they are much less. Only time will tell.

Because of this uncertainty, we are still a long way from knowing if the strange octonions are of fundamental importance in understanding the world we see around us or merely a piece of beautiful mathematics. Of course, mathematical beauty is a worthy end in itself, but it would be even more delightful if the octonions turned out to be built into the fabric of nature. As the story of the complex numbers and countless other mathematical developments demonstrates, it would hardly be the first time that purely mathematical inventions later provided precisely the tools that physicists need.

More to Explore

An Imaginary Tale: The Story of the Square Root of −1. Paul J. Nahin, Princeton University Press, 1998.

The Octonions. John C. Baez in *Bulletin of the American Mathematical Society*, Vol. 39, pp. 145–205; 2002. Paper and additional bibliography at http://math.ucr.edu/home/baez/octonions

Ubiquitous Octonions. Helen Joyce in *Plus Magazine*, Vol. 33: January 2005. http://plus.maths .org/content/33

Mathematics Meets Photography: The Viewable Sphere

DAVID SWART AND BRUCE TORRENCE

Right now, without moving from your seat or from where you are standing, look at your surroundings. Not just left and right but all the way around to whatever is directly behind you too. Look at the ceiling or the sky. Look at the floor below, or maybe it's a desk or a laptop below your nose. What you can see from your single point of view is a *viewable sphere*. Perhaps it helps to picture an imaginary sphere surrounding your head with imagery printed on it that matches your surroundings.

Of course, we need to refine this idea slightly to make it precise. A sphere has a single center, and you probably have two eyes. So stay very still, shut one eye, and imagine a sphere centered at the optical center of your open eye—the point where the light rays converge on their way to your retina. The radius of this imaginary sphere is not so important; let's just make it large enough so that your entire head is on the inside, and small enough so that it lies between you and every object within sight. Then for each point in the scene around you, even those behind you, the line segment from that point to the center of your open eye intersects the sphere in a unique point. In this way, the scene around you can be painted, or at least projected conceptually, to create a well-defined viewable sphere.

With the advent of digital photography and continuously improving photo-stitching software, panoramas that capture the entire viewable sphere have become more and more commonplace.

To shoot an all-around panorama, a photographer takes a series of overlapping photographs in every direction from the exact same point in space—the optical center of the lens. It helps to have a wide-angle lens and a specialized tripod for this purpose. The photographer then imports these photos into software that can stitch them together into a seamless panorama.

These viewable spheres can be explored interactively. You might be familiar with applications such as Google Street View or panorama viewers that give online virtual walk-throughs of new homes, hotel rooms, or cruise ships. Others have created actual physical spheres with the panorama printed on them. Dick Termes, for instance, paints viewable spheres in what he calls "six-point perspective" because the viewable sphere contains all six vanishing points (two for each of three orthogonal directions).

However, a flat image is in many ways more accessible and practical than these computer-dependent applications. If we can find a mathematical mapping, or *projection*, that can map points from a sphere to a plane, then these viewable sphere panoramas can be printed and shared, hung on the wall, shown on a flat computer display, or included in a book or in the magazine *Math Horizons*. We need a function that takes points on the viewable sphere as input and outputs locations in the two-dimensional plane.

Luckily, cartographers and astronomers have been addressing this problem for millennia. What keeps cartography both interesting and difficult is a theorem by Riemann, which states that any mapping from a sphere to a plane introduces some sort of distortion. Cartographers are naturally concerned with which sorts of distortions a projection introduces and whether a projection is suitable for the map-user's purpose. For instance, a specific projection might fail to preserve area, distances, angles, orientation, or any combination of these.

To get an idea of how projections are created or chosen for a purpose, consider an example. The *Mercator projection* is well suited for navigation because a straight line on a Mercator map corresponds to a path a ship would take while keeping its compass bearing constant. However, the Mercator map is not ideal if equal area is a concern because of the large changes in scale, especially near the poles.

Let's consider the *equirectangular projection* as a candidate to map viewable spheres. Even by cartography standards, it is old, attributed to Marinus of Tyre, circa AD 100. In this projection, the *meridians* (lines of longitude) are set as equally spaced vertical lines, and the *parallels* (lines of latitude) are equally spaced horizontal lines. For an equirectangular projection, you map a point from a sphere to a plane by simply renaming longitude and latitude to x and y, respectively. The left and right edges correspond to a single meridian line, while the top and bottom edges correspond to the North Pole and South Pole, respectively.

The format yields an image with a 2:1 aspect ratio because it takes 360 degrees to go all the way around the world longitudinally, but only 180 degrees to go from the North Pole to the South Pole. The equirect-angular projection is so straightforward and useful that it has become the de facto standard native format for storing digital versions of view-able spheres.

A word of warning: In addition to cartography jargon such as meridi-ans, parallels, equators, and poles, our discussion may include panorama terminology such as *zenith* (the point on the viewable sphere directly above the observer), *nadir* (the point below), and the *horizon* line.

Looking at the equirectangular projection in Figure 1, we see that everything is oriented well (things that are pointed upward in the vis-ible sphere are pointing upward in the equirectangular projection). However, as we can see, the equirectangular projection is not very well suited for viewing the parts of the panoramas near the zenith or the nadir. There is a rather nasty horizontal stretching at the top and the bottom of the projection.

Granted, these parts often comprise uninteresting flooring, grass, sky, or ceiling. But the features that are there are unrecognizable and can detract from the panorama's aesthetics. We want to look for projections that produce panoramas where no features are skewed or stretched. Do such projections exist? Under a reasonable interpretation of the phrase "no features are skewed or stretched," the answer is a resounding yes!

Figure 1. Equirectangular projection. Photo by Sébastien Pérez-Duarte.

Conformal Mappings

When looking at different attributes of projections, one property seems to be better suited for photographic content than others: *conformality*. Mathematically speaking, conformal mappings are mappings that preserve angles at a local level. For instance, if two curves meet at, say, a 45-degree angle on the viewable sphere, then their images in the plane under a conformal mapping also meet at 45 degrees. To be conformal, a mapping must preserve every angle on the viewable sphere.

Qualitatively speaking, conformal maps ensure that the imagery does not get sheared or skewed; there is no squishing and stretching of the sort that we see near the top and bottom of an equirectangular projection. (While delightfully simple, equirectangular projections are *not* conformal.) As a consequence of preserving angles, it can be shown that any squishing or stretching that does happen under a conformal mapping occurs in all directions uniformly. This phenomenon ensures that small features retain their shapes, even though we can expect some variations in scale and orientation. In other words, small-scale details keep their overall shape and appearance, but large-scale shapes might grow, shrink, bend, or become distorted in some manner. This change of scale is not as problematic for photographic content as it is for maps, perhaps because your eye tends to accept larger and smaller details as closer and farther away, respectively.

The conformal *stereographic projection* is about as old as the equirectangular projection. It is attributed to Ptolemy in the second century AD. It is one of the most popular projections for viewable spheres. If we imagine the viewable sphere itself as a translucent ball resting on a white floor, we could place a lightbulb at the zenith of the sphere and then look at the floor. The imagery on the floor is the stereographic projection of the sphere (Figure 2). This mathematical function lives up to its name as an actual projection.

The stereographic projection of a viewable sphere is striking. Everything below the horizon is transformed to the interior of a circular region reminiscent of the little planet from Saint-Exupéry's *Little Prince*. So it is no surprise that these projections have been dubbed "Little Planets" (Figure 3).

In a stereographic projection, the nadir of the panorama is moved to the center of this little planet sphere. Because the nadir is so prominent,

Figure 2. Stereographic projection transforms the spherical image into a flat one.

Figure 3. Stereographic projection produces a "little planet" effect. Photo by Sébastien Pérez-Duarte.

it is not uncommon to see panoramas that are taken from a point directly above a decorative floor element.

The horizon, which forms the equator of the viewable sphere, gets projected onto a circle. Everything below the horizon lands inside the circle, and everything above the horizon lands outside it. The photographic details at the horizon are such that the horizontal plane of the ground underfoot is parallel to the viewer's line of sight. So when this

Figure 4. Stereographic projection of a viewable sphere centered above an infinite checkerboard plane.

horizon is mapped to a circle, it gives the appearance of being the edge of a sphere viewed head on, as you can see in Figure 4.

Lines to Circles

Stereographic projection can be better understood by seeing what it does to something as simple as a line in the original scene. What happens to the edge of a sidewalk or building, or to the edges of the book that you are now reading?

Recall that we first project the three-dimensional scene onto the viewable sphere. We then apply stereographic projection to the spherical image. We need to follow a line in the original scene through these two stages. For the first stage, start with any line in the scene that does

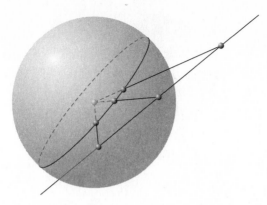

Figure 5. A line in the scene becomes a great circle on the viewable sphere.

not contain the center O of the viewable sphere. It will be projected to a *great circle* on the viewable sphere, as shown in Figure 5. To see why, simply note that the line together with the center O determine a plane. Since this plane passes through the center of the viewable sphere, it intersects the sphere in a great circle.

For the second stage, we are left with the task of determining what happens to a great circle under stereographic projection. We see that there are two distinct cases: those great circles on the sphere that pass through the zenith and those that do not.

The first case is simple. Any great circle containing the zenith must also contain the nadir. Such circles are meridians; as such, they correspond to *lines* under stereographic projection. Every such line passes through the nadir at the center of the final image, and we could reasonably call them *radial* lines. Since *vertical* lines in the original three-dimensional scene are portions of meridians on the viewable sphere, we conclude that vertical lines in the three-dimensional scene map to radial lines in the stereographic projection. More generally, any line meeting the horizon at a right angle corresponds to a meridian on the viewable sphere, and thus maps to a radial line in the stereographic projection. Can you see the radial lines in Figure 4?

The second case is slightly less simple. It can be shown that any circle (great or otherwise) on the surface of the sphere, that does not pass through the zenith, maps under stereographic projection to another *circle*. (See Figure 6, and consult the Needham reference in the Further

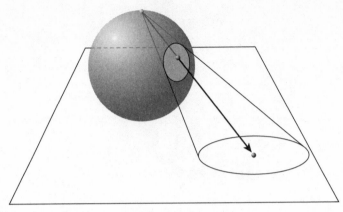

Figure 6. Stereographic projection maps circles that do not contain the zenith to circles.

Reading section for details.) This case implies that lines in the original scene that are not meridians on the viewable sphere map stereographically to *circles*. Lines that intersect the horizon at almost a right angle (great circles that are almost meridians on the viewable sphere) map stereographically to large circles. Lines that are closer to horizontal map to smaller circles. The smallest circle in the stereographic projection that corresponds to a line in the original scene is the horizon circle; no great circle on the sphere projects stereographically to something smaller.

So we have answered the question of what happens to linear features in the three-dimensional scene. Every straight line in the original scene becomes, under stereographic projection, either a portion of a radial line or some portion of a circle, according to whether or not the original line corresponds to a meridian on the viewable sphere. Moreover, it is easy to see that any sphere projects in a circle on the viewable sphere and maps to a circle via stereographic projection.

Now close one eye again, and look for lines in the scene around you. Which of them correspond to meridians on the viewable sphere, and so become radial lines in the stereographic projection, and which are skew to the horizon, and so become circles instead? Can you imagine what the little planet version of your current surroundings would look like?

Sphere Tipping

The operation of stereographic projection can be generalized in a manner that is both simple and extremely useful. If we think of the viewable sphere as a translucent orb resting on a white projection plane, a bright light at the North Pole projects the imagery from the sphere to the plane. The result is a striking image that looks like a little planet floating in the sky (Figure 7). We might ask what the panorama would look like if we were to tilt the ball (while keeping the lightbulb directly above it, opposite the floor). It turns out this sphere-tilting operation has two useful properties: It preserves conformality, and circles that were in the original stereographic projection, before tilting, still appear

Figure 7. Stereographic projection. Photo by David Swart.

as circles after the transformation. In particular, the circular image of the horizon remains circular.

This tilting operation is an excellent compositional tool, for it allows the panoramic photographer to emphasize the interesting parts and deemphasize the boring bits. Rotating the image of, say, a building to and from that lightbulb at the top makes the building appear larger and smaller in the projection on the floor, in much the same way that the shadow of your hand in front of a lightbulb would grow and shrink (Figure 8).

Figure 8. Tilting the sphere slightly before stereographic projection. Photo by David Swart.

Thus, we have a class of transformations that keeps little planets looking like little planets, at least as long as we don't tilt the sphere too far.

You may recall that a great circle on the sphere that passes through the North Pole (a meridian circle) maps stereographically to a line through the center of the projected image. If we tilt the viewable sphere by 90 degrees, the horizon circle from the original scene becomes a meridian and so projects to a line. Tilting by a little more or a little less than 90 degrees maps the horizon to a *very* large circle—it appears to be almost a straight line when projected stereographically. The sky lies on one side and the ground on the other, much like we see in "ordinary" photographs. But in such a projection we enjoy almost a full 360-degree field of view along the horizon (impossible in an ordinary photo!), and this fact implies that there must be some serious distortion. See Figure 9 for an example.

And if we tilt the viewable sphere a full 180 degrees, so that it is entirely upside down before projecting, then the horizon maps to the same circle as in the original projection. But now it is a circle with the sky on the *inside*—resulting in a fantastical-looking tunnel world (Figure 10).

Figure 9. Tilting the sphere almost 90 degrees before stereographic projection. Photo by Bruce Torrence.

Figure 10. Tilting the sphere upside down before stereographic projection. Photo by David Swart.

The Riemann Sphere and the Complex Plane

While stereographic projection has been used in cartography for centuries, in the 19th century Bernhard Riemann used it to gain a critical insight into the complex numbers. If one associates the projection plane beneath the viewable sphere with the complex plane—so that the sphere has radius 1 and rests precisely on the origin—then stereographic projection provides a one-to-one correspondence between the complex numbers and the points on the sphere, except for the North Pole. Said another way, if a single point were added to the set of complex numbers, this "extended complex plane" could be placed in perfect one-to-one correspondence with the points on a sphere via stereographic projection. The extra point is called the *point at infinity*, and the sphere in this context is known as the Riemann sphere.

It so happens that the field of complex analysis, from Riemann's time to the present, has produced a cornucopia of conformal mappings from the extended complex plane to itself. This gives the panoramic photographer a deep reservoir of possibilities for producing flat images from the viewable sphere. The new possibilities come about by identifying the viewable sphere with the extended complex plane via stereographic projection and then applying conformal mappings from the extended complex plane to itself to obtain new images.

In fact, the simple sphere-tilting operation described earlier can be described this way: It can be thought of as a *Möbius transformation*. Mathematically, a Möbius transformation is a function on the extended complex plane having the form

$$f(z) = \frac{\alpha z + \beta}{\gamma z + \delta},$$

where α, β, γ, and δ are complex constants. If the photographer applies stereographic projection to the nontilted sphere and identifies each pixel location (x, y) in the projected image with the complex number $z = x + iy$, then $f(z)$ gives the location of that pixel under a tilt-then-project operation. The values of the four constants determine what the tilt is, as well as accounting for zooming in or out, and possibly recentering the image. The short video "Möbius Transformations Revealed" (Arnold and Rogness) gives a visual explanation of how this works. Chapter 3 of Tristan Needham's book *Visual Complex Analysis* provides the mathematical details.

A charming special case is the inversion function $f(z) = 1/z$. Note that in the extended complex plane, $z = 0$ is mapped to the point at infinity, and vice versa, so this function interchanges the north and south poles of the viewable sphere. It corresponds to a 180-degree tilting of the viewable sphere (through the axis parallel to the real axis in the complex plane below) and produces the tunnel effect discussed earlier.

Many other complex functions are "mostly" conformal—if you're willing to avoid or tolerate a few isolated *singularities* (points where conformality breaks down). You may also have to contend with a

Figure 11. The function $f(z) = z^3$. Photo by Sébastien Pérez-Duarte.

many-to-one function. For instance, the power function $f(z) = z^n$ for any integer $n > 1$ is conformal except at 0 and infinity. Visually, this function opens and closes the imagery radially around origin like a folding fan, a total of n times, as in Figure 11.

The exponential function $f(z) = e^z$ turns horizontal lines in the original stereographic projection to radial lines through the origin, and vertical lines into concentric circles. It tends to produce perplexing imagery, even in this warped arena. Figure 12 shows the effect of the function $f(z) = (z + 1/z)$ on a stereographic panorama.

Mercator and Beyond

The stereographic projection is not the only way we can conformally project the viewable sphere onto the plane. In the 16th century, Gerardus Mercator developed the famous projection named after him. The Mercator projection is similar to the equirectangular, but it is stretched vertically in a special way. It is the only conformal projection that keeps its lines of latitude horizontal and its lines of longitude vertical, although it does have the odd side effect of making the polar regions disproportionately large, and placing the North and the South Poles infinitely far away. As with physical maps, panoramas mapped with the Mercator projection benefit from a judicious cropping:

Figure 12. The function $f(z) = (z + 1/z)$. Photo by David Swart.

We rendered a pic in Mercator
and returned just a little while later.
    We got a surprise
    seeing the file size.
We'd included the zenith and nadir!

Practical limitations notwithstanding, we can still apply complex functions to Mercator projections to create new conformal projections.

Figure 13. Lagrange projection. Photo by Bruce Torrence.

Let's consider one interesting case that came from Lambert in 1772, which in an odd twist is named after Lagrange (who provided a generalization in 1779). We first apply the relatively simple $f(z) = z/2$. This mapping takes the infinitely long ribbon that is the Mercator projection and shrinks it down to half its size: an infinitely long ribbon half the width of the original. Applying the *inverse* Mercator projection conformally puts this half ribbon back onto the sphere—but now it is on only half of the sphere. If you rotate so that this half of the sphere is the lower half and apply stereographic projection, the image is mapped onto a disk. This feat is remarkable; the entire sphere is conformally mapped onto a disk (Figure 13).

A curious thing about conformal mappings is that mathematicians seem to customarily define conformal mappings between specified

Figure 14. A tiling based on the Peirce quincuncial projection. Photo by Sébastien Pérez-Duarte.

shapes and the unit disk. And lucky for us, stereographically projected hemispheres and Lagrange projections are disks, and thus great candidates for many conformal mappings. These mappings provide a great opportunity for novel projections. For instance, in the late nineteenth and early twentieth centuries, complex analysis had advanced enough to allow mathematicians to conformally map the unit disk to a square. Notably, the *Peirce quincuncial projection* has the entire viewable sphere conformally mapped onto a tileable square, which allows for some space-filling panoramas (Figure 14).

But You Don't Have to Take Our Word for It

Although we have by no means presented an exhaustive list of the various conformal projections to which spherical panoramas have been subjected, perhaps we've provided enough to encourage you to try some of them yourself. Photo-stitching panorama tools such as Hugin can output some of the projections we've talked about. From there, you can use the Mathmap plug-in for the GIMP, Flaming Pear Software's *Flexify* filter for Photoshop, *Mathematica*'s image-processing capabilities, or your favorite programming language. A few minutes of poking your nose into a complex analysis textbook is likely to yield some equations for some fascinating conformal maps. Or just venture out on your own and see what twisted ideas you can come up with yourself.

Further Reading

Examples of panoramic imagery abound at Flickr.com; search for "stereographic" or "equirectangular."

Douglas Arnold and Jonathan Rogness, "Möbius Transformations Revealed" (a beautiful short video); ima.umn.edu/~arnold/moebius/.

Carlos Furuti maintains an excellent website cataloging properties of various cartographic projections: visit http://www.progonos.com/furuti/MapProj/CartIndex/cartIndex.html.

The geometry of stereographic projection, and in particular the fact that circles map to circles, is explained beautifully in Chapter 3 of Tristan Needham's book, *Visual Complex Analysis* (Oxford University Press, 1999).

If you want to try making your own panoramas, the (free, open source) Hugin stitching software is a good place to start. Smartphone apps such as Autostitch and Photosynth are gaining popularity and are simple to use.

Dick Termes has a website that shows images drawn in his six-point perspective style, http://termespheres.com/

Dancing Mathematics and the Mathematics of Dance

SARAH-MARIE BELCASTRO AND KARL SCHAFFER

If you're not a dancer—and even if you are—you may be wondering how on earth mathematics and dance are related. We're not talking about social or folk dances such as contra dance—a popular dance form among mathematicians—but about dance as an artistic endeavor. There are superficial links such as counting steps or noticing shapes, but also deeper connections, such as mathematical concepts arising naturally in dance, mathematics inspiring dance, or using mathematics to solve choreographic problems.

We are both mathematicians and both dancers. For about 20 years, we have independently thought about ways in which dance and math intermingle. Our interests intersect in mathematics through graph theory and in dance via the field of modern dance. Modern dance encompasses a variety of nonclassical forms that originated mainly in the United States and, despite its name, is almost 100 years old. sarah-marie's dance background is primarily in ballet and modern dance, though she also has some experience in tap, jazz, West African, Silvestre technique (Brazilian modern), and ballroom dance. Her Ph.D. is in algebraic geometry, but she now does topological graph theory research. Karl's dance work has been primarily in modern or contemporary dance, though he has studied and performed tap, Bharatya Natyam (South Indian classical dance), folk dance, and other forms. He has co-directed a modern dance company with Erik Stern for about 20 years and teaches dance on a freelance basis. Together, separately, and with puzzle designer Scott Kim, Karl and Erik have created—among other things—a series of concerts around mathematical themes. Karl's Ph.D. and research are in graph theory.

Mathematics within Dance

Many mathematical ideas pervade dance and, we would argue, are intrinsic to dance. For example, we divide music into counts and use counting to mark the times at which movements are done. As mathematicians, we might not think of this as deeply mathematical, but many dancers and musicians do see this as the presence of mathematics, and complex patterns inevitably arise when dancers play with rhythm.

Each dance tradition has its own characteristic way of using mathematical concepts. For example, classical western ballet and Bharatya Natyam both use a strong sense of line. However, as Karl's teacher Kathryn Kunhiramen pointed out to him, in Bharatya Natyam the dancer's lines end—they are cut off by abstract or representational *mudras* made by flexing the hands. This situates the dancer "in the world," rather than extending beyond it into the "world of the gods." In contrast, the ballet dancer's lines extend toward infinity, symbolizing an endless extension over the natural world.

Symmetry is important in dance, as is pattern. These elements have local and global forms. Movements of a single (local) body can be symmetrical—for example, bilaterally or with a glide reflection along the time axis—and a dancer can execute a pattern of movements that repeats. On the global scale, there are symmetries between groups of people moving, as when the movements of one set of bodies are symmetrical to the movements of another set of bodies, or when dancers are placed symmetrically on the stage. There are many more possibilities for pattern on the global scale than there are on a local (single-person) scale. Some dance forms regularly use particular symmetries. For example, in the flamenco duet, two dancers circle each other closely with 180-degree rotational symmetry, creating a dynamic of intimate opposition. But it is also possible to find mathematical structure even in the seemingly simple business of having one dancer move in a straight line. John Conway created a "hop-step-jump" terminology to describe the seven linear patterns for ambulation shown in Figure 1. *The Symmetries of Things*, by Conway, Burgiel, and Goodman-Strauss, is a good reference [1].

We can also look at how symmetries interact with each other [3]. Using just four symmetries—translation, mirror reflection, 180-degree rotation, and glide reflection—we can create what is called the Klein

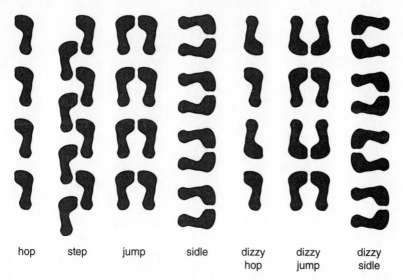

| hop | step | jump | sidle | dizzy hop | dizzy jump | dizzy sidle |

Figure 1. Conway's "hop-step-jump" symmetries.

Try this: Attempt to execute each of the patterns in Figure 1 repeatedly. For the "dizzy" patterns, you'll have to do (at least!) a 180-degree turn for each repetition and decide whether to turn to the right or the left. Which moves are the most difficult? The easiest?

four group, or $Z_2 \oplus Z_2$, by combining the symmetries pairwise. For example, in Figure 2, a mirror reflection of the first dancer on the left gives the position of the second dancer in the middle; a 180-degree rotation of the middle dancer then gives the position of the third dancer on the right, which is the same result as if we had simply applied a glide reflection to the first dancer. A similar thing occurs when we compose any pair of symmetries from the original list—the result is always one of the four symmetries.

In practice, dancers use other kinds of symmetries in three dimensions, including helical symmetry and central inversion (the arms inverting to the legs through a central point in the solar plexus). Dancers and choreographers might not use this terminology, but they do have

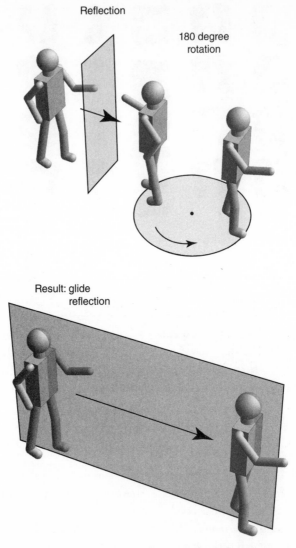

Figure 2. Composing a mirror reflection with a rotation yields a glide reflection.

to be adept at performing symmetries. For example, a choreographer might say to a dancer, "Do the same phrase but on the other side (reversing right and left), facing the opposite way, and two counts behind." There are symmetries in time and space that dancers do name: (a) unison (a set of movements performed at the same time); (b) canon (a set

Try this: Figure 2 shows how the composition of a mirror reflection followed by 180-degree rotation is a glide reflection. A mostly empty table is given below—fill it in with compositions of translation (T), glide reflection (G), 180-degree rotation (R), and mirror reflection (M), by acting them out or, we might say, by allowing the symmetries to act on the positions of your body. For clarity, do not start from a position that has bilateral symmetry within your body. This requirement makes it difficult to distinguish which of the four symmetries is being performed. The table you get when you are done is really the multiplication table for the group . Can you explain the patterns in the table in terms of the movements they represent [3]?

	T	*G*	*R*	*M*
T				
G				
R				*G*
M				

Try this: Using three people, position your bodies to demonstrate a composition of two symmetries as in the top row of Figure 2. Now attempt moving slowly while keeping the symmetries intact. You might then sequence several such movement phrases together to make a longer dance sequence, use more than three dancers, or try setting your sequence to music.

of movements offset in time from each other); (c) inversion (movements performed in reverse sequence); and (d) retrograde (each movement is reversed, in addition to the sequence being reversed).

Try this: Make up a sequence of four movements that includes a leap, then attempt to retrograde it. Notice that retrograding a leap is particularly challenging!

Beyond Symmetry

There are other mathematical ideas involved in dance. One can pay attention to the geometry of the moving body; to the topology of links between dancers, for example, during lifts; to the spatial paths of the dancers; to the interaction of dancers with simple props such as ropes or poles (or mathematical props such as giant tangram pieces); or even to conceptual problems like game theory in a struggle between two dancers. These notions have all been used in the authors' choreography.

Explicit mathematics also shows up in the history of modern dance, specifically in the work of Rudolf Laban [2], a European dance theorist from the early twentieth century (Figure 3). Describing the mathematics in his work would take an entire article, so here we'll give just a few examples. Laban described the directions, types, and qualities of movements within the *kinesphere*—that's the sphere with radius "arm" or "leg" centered on the navel—using regular polyhedra. The usual coordinate axes correspond to movement qualities, with light/strong for the z-axis, directed/undirected for the x-axis, and slow/quick for the y-axis. Applying this convention, the corners of a cube then correspond to the eight movement qualities that arise from taking one of the two choices for each of the three coordinate directions (Figure 4):

press (slow/strong/directed),
wring (slow/strong/undirected),
glide (slow/light/directed),
float (slow/light/undirected),
slash (quick/strong/undirected),
punch (quick/strong/directed),
dab (quick/light/directed), and
flick (quick/light/undirected).

Figure 3. Modern dance theorist
Rudolf Laban sitting amid the
polyhedra that fascinated him.

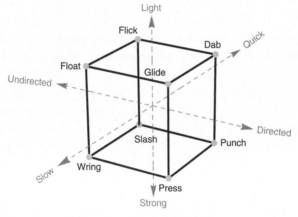

Figure 4. Each corner of Laban's cube corresponds to a particular movement
quality determined by the three corresponding coordinate directions.

Laban's system has evolved into an exercise in which the dancer traces
out movement qualities in succession while directing effort toward the
corresponding cube vertices.

    Laban used polyhedra in other aspects of his theoretical work. He
asserted that extending the limbs toward the six vertices of an octahedron (forward, back, right, left, up, and down) leads to stability,
whereas reaching toward the vertices of a dual cube (high back right
and left, low back right and left, high forward right and left, low forward left and right) initiates motion.

Conscious Choreographic Connections

Each of the authors has created mathematical dances, some inspired by mathematics, some inspiring mathematics, and some where the mathematics and the dance are so deeply intertwined that it would be difficult to say which impulse started the creative process. Although it's difficult to describe dance in text, we'll attempt it here in order to give examples of the interplay between mathematics and choreography.

In 1990, Karl and his frequent collaborator Erik Stern created *Dr. Schaffer and Mr. Stern, Two Guys Dancing About Math*, a show they've performed more than 500 times throughout North America. In the performance, the instinctual Mr. Stern and the rational Dr. Schaffer mix arguments about mathematics in daily life with dances exploring the physics of motion of a basketball, the complex rhythms of tap dance, vaudevillian handshake permutations, and flyswatters of ever-increasing sizes.

In 1999, sarah-marie choreographed *Crystalline Meringue*, a ballet trio. Much of the structure arises from actions of the symmetric group on three dancers. For example, the three dancers often rotate positions and then separate into different pairings of two dancers doing one type of movement and a third dancer doing a second type of movement. True to the group structure, the position of the lone dancer is more important than which dancer she is. In a later piece for four dancers called *Swirly Suite I*, sarah-marie exhibited the binomial coefficient 4-choose-2 by pairing each dancer with every other dancer. She also mapped the movements "grapevine" and "balloncé" to 0 and 1 respectively, so that pairs of dancers corresponded to elements of the Cartesian product $Z_2 \times Z_2$.

In 1995, Karl, Scott Kim, and Barbara Susco created *Trio for Six*. It began with a desire to do something involving "finger geometry" (Figure 5), but it evolved during creation and eventually included illusions and finger puppetry.

Karl often uses simple props as giant math manipulatives, especially when in collaboration with Erik Stern and Scott Kim [3]. He recently created *Fragments*, a dance involving oversized tangram pieces that is a serious exploration of the fragmentation and destruction of war. He has explored polyhedra in dance using various props: PVC pipes, loops of string, and even fingers. In one dance he used linked and glow-painted

Figure 5. A finger tetrahedron, invented by Scott Kim.

Try this: Find a way to form a cube with a partner, each person using the thumb and first two fingers of each hand.

Try this: For a related puzzle, how can the "hexastar" shown in figure 6 be folded at the vertices to form a cube? An octahedron? Two tetrahedra?

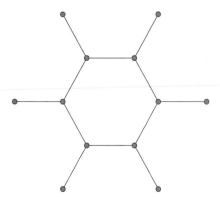

Figure 6. The "hexastar."

PVC pipes to make both polyhedra and whimsical shapes. This is an example of how mathematical ideas have led to dance movement that then led back to a mathematical question. In this case, the question is: What is the largest linkage of equal-length PVC pipes (i.e., what is the unit-edge-length graph with the largest number of vertices), multiple copies of which can decompose the skeletons of all five of the Platonic solids?

The answer is the graph ∘∘⅃∘∘, a proof of which will appear in a forthcoming paper.

How Does Mathematics Sound?

Rhythm is an important part of any dance, and some dance forms, such as tap and clogging, use sound extensively as well. Figure 8 shows a seven-pointed star—a polygon denoted by the Schläfli symbol {7/5}. The 7 in this notation refers to the seven vertices around the circumference, and the 5 refers to how they are connected. Starting with a

Figure 7. sarah-marie's topological graph theory class at the 2004 Hampshire College Summer Studies in Mathematics program, dancing a Petersen graph.

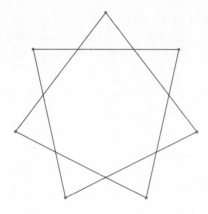

Figure 8. A seven-pointed star with Schläfli symbol $\{7/5\}$.

vertex, we connect it to the fifth vertex counting counterclockwise around the outside and continue until we end up back at the vertex where we started. In this case, all the vertices are connected before we return to the starting vertex.

This pattern can be demonstrated by having one person clap a five-beat rhythm and another clap a seven-beat rhythm with the same tempo, and at the same time. Both people clap loud sounds on beat 1 of each phrase. This is a way to create interesting syncopations in which the accented beats don't fall in the expected place. In this case, they fall at every possible location within the 7-phrase before the entire pattern repeats after the LCM(5,7) number of beats [3].

It's not hard to find a piece of music that repeats its rhythmic pattern every eight beats, with a strong accent on the first beat, but what is challenging is to clap on every third beat, beginning with the first (Figure 9). If we draw the eight-pointed star, which we notate $\{8/3\}$ in the spirit of the previous example, then, as before, the claps occur at every one of the eight different beats as we move through the LCM(8,3) = 24 beats, and the star is again one continuous strand.

Of course, the claps can be replaced by tap shoes. In *Lost Souls* from 2000, Dr. Schaffer and Mr. Stern tap a seven-beat phrase against an

X	2	3	4	5	6	7	8	X	2	3	4	5	6	7	8	X	2	3	4	5	6	7	8	X
X	2	3	X	5	6	X	8	1	X	3	4	X	6	7	X	1	2	X	4	5	X	7	8	X

Figure 9. Clapping on every third beat against an eight-beat count produces a single clap on each of the eight beats before repeating.

eight-beat score. They also play with other aspects of rhythm, alternating sections of the phrase between them, using canon, and rearranging the direction of travel, while commenting drolly on the relative merits of sevens versus eights.

Try this: With a partner, demonstrate the $\{8/3\}$ pattern just discussed by having one person clap the eight-beat rhythm and the other clap the three-beat rhythm, each clapping loudly on the first beat of each cycle. Or try it yourself, clapping the eight-beat rhythm while stomping every third beat.

How many k in the set $\{1, 2, 3, \ldots, 12\}$ make the $\{12/k\}$ star polygon one continuous strand that hits every vertex of the 12-gon? What about for n in general?

Math for More Practical Purposes

Sometimes as choreographers we call on mathematics for its simple usefulness. In sarah-marie's *Swirly Suite II* and *III*, the entrances were challenging. In Part *II*, each dancer had to follow the same path, which crossed the stage and then wrapped around its perimeter in an oval; the dancers also had to be evenly spaced in space and in time, not crash into each other, and end up in a circle. Part *III* of the dance required the dancers to appear evenly along the diagonal of the stage, but in random order. This requirement was also a challenge for the dancers because there was no beat in the music and yet the dancers had to count seconds before entering. Later, the dancers were in a circle but needed to leave the stage imperceptibly. The solution was to have the circle open into a line segment, and then shrink to a point as the dancers moved toward the center of the segment, while every two counts the middlemost dancer exited into shadows.

In one of Karl's most recent pieces, two dancers used game theory in a simple way, playing "*The Atom Bomb Game*." In this game, the dancers each secretly choose a number less than 100, the larger number winning, except that if the numbers add to 100 or more, they both might die. It's a game of brinksmanship, except that one of the dancers doesn't want to play! The dance is a way of using mathematics as one of several metaphors with which to expose serious issues, albeit in a humorous way—namely, the strange ways we talk about nuclear Armageddon, or avoid talking about it.

Even more recently, the two of us performed a very short new work demonstrating the classic result from geometry that the angles of a Euclidean triangle sum to 180 degrees. The idea is that one can physically sum the internal angles of a triangle by traveling along each side in succession and turning through the internal angle at each vertex. Upon arriving at the beginning point of the triangle, the dancer is facing in the opposite direction of his or her starting direction. Once Karl had the idea, we each independently choreographed such a dance, and of course they were stylistically different. At the 2008 Joint Mathematics Meetings, we performed the dances separately, and then simultaneously. This idea might be extended to other surfaces by using the Gauss-Bonnet formula, though the aesthetics of such a dance are somewhat elusive.

Try this: Walk the borders of a square or pentagon forward and backward, turning through the internal angles [4]. How many rotations will you perform?

To be clear, we don't view dance entirely through the lens of mathematics—or vice versa. Even when a dance has a strong mathematical element, we let the dance take on a life of its own. Dance is many things, sometimes all at the same time: artistic expression, ceremony, social interaction, political protest, expression of sexuality, a form of athletic competition, physical exercise, theater, psychological catharsis, community event. And sometimes it's the interplay of these elements with mathematics that engages us as artists—and, we hope, engages the audience as well.

Solution to Composition of Symmetries "Try This" Challenge

	T	*G*	*R*	*M*
T	*T*	*G*	*R*	*M*
G	*G*	*T*	*M*	*R*
R	*R*	*M*	*T*	*G*
M	*M*	*R*	*G*	*T*

Further Reading

[1] John Conway, Heidi Burgiel, and Chaim Goodman-Strauss, *The Symmetries of Things* (A. K. Peters, 2008).

[2] Rudolf Laban with Lisa Ullmann, *The Language of Movement: A Guidebook to Choreutics* (Plays, Inc., 1974).

[3] Karl Schaffer, Erik Stern, and Scott Kim, *Math Dance with Dr. Schaffer and Mr. Stern* (MoveSpeakSpin, www.movespeakspin.org, 2001).

[4] Karl Schaffer, "Harmonious Dances," in *Bridges Banff Proceedings 2009*. (MathArtFun.com).

Can One Hear the Sound of a Theorem?

Rob Schneiderman

Mathematics and music have been intertwined in a long-running drama that stretches back to ancient times and has featured contributions from many great minds, including Pythagoras, Euclid, Mersenne, Descartes, Galileo, Euler, Helmholtz, and many others (e.g., [1]). Applications of mathematics to music continue to develop in today's digital world, which also supports active communities of musicologists and experimental composers who examine music methodically, often using mathematical elements. In light of the recent wave of musico-mathematical books, blogs, journals, and even articles in the *Notices of the American Mathematical Society* (*Notices*), this multifaceted side of the mathematical world deserves reexamination. Although the scrutiny given here reveals many problems posing as solutions, some promising prospects also emerge, and positive turns in the plot may yet unfold, especially when viewed from a novel educational angle, which I describe here.

From the mathematician's perspective, besides providing a bounty of physical applications, the search for relationships between music and mathematics should serve both as a philosophical reflection pool and as a portal to an engagement of the general public with mathematics. But the view is often obstructed by the unwitting entanglement of several distinct lines of thought. It is not uncommon for commentary on music and mathematics to bounce among the physics of sound, theoretical analysis of music, and metaphorical prose. Although each approach has its strengths and weaknesses, unjustified juxtapositions can serve to cloud the big picture by masquerading as implicit unifications of unresolved key issues or by appearing to support pseudoscientific arguments. For example, [4] and [5] exposit useful mathematical techniques in the setting of digital audio processing, which are then associated with flawed musical analysis and exaggerated conclusions. A historical article on the mathematics of fretting a guitar in [6] is presented side-by-side with musical numerology in a collection whose introduction enthusiastically includes as evidence of connections between mathematics and music the "ordering by number" of Bach's Goldberg variations!

As one who came to mathematics after a career as a professional musician, I offer here a personal viewpoint in hopes that it will provide a helpful framework for unwinding the current strands of a fascinatingly elusive subject. This essay argues that whereas mathematics provides satisfying analyses of sound and useful parameterizations of musical choices, deeper scientific relationships between mathematics and music remain largely beyond reach. But the adoption of a more metaphorical point of view uncovers support for a return of music and mathematics to a quadrivium-like partnership in education that is based on a common strength of intrinsic structure.

The goal here is not to give a survey of the present state of musico-mathematical affairs but rather to highlight a representative sample of points that seem to be overlooked or underappreciated in the current general discourse. Of course, personal taste enters into any discussion of music, and many issues raised below are subject to differing interpretations. The arguments are mostly critical because such objections seem to have had trouble finding their way into print, but I support many aspects of even the approaches criticized here and hope to clarify and stimulate the ongoing dialogue. It is in the interest of the mathematics community to engage in and be aware of the development of interdisciplinary work in all directions.

The body of this article is roughly divided into the subtopics of *the science of sound, analysis of music,* and *metaphorical comparisons.*

The Science of Sound

Much solid and fascinating mathematical work, classical and ongoing, is related to musical sound, including instrument design, acoustics, and audio processing, among many interrelated topics. Applications of mathematics are readily apparent in the modern recording studio, where the signal of digitally recorded instruments (both electric and acoustic) is routinely manipulated in a wide variety of ways, including the independent adjustment of tempo and pitch of individual voices, as well as the elimination of ambient noise and the creation of audio effects. Fourier theory plays a central role throughout these settings, essentially because of the periodic nature of musical sound waves and the graded elasticity of our ears' basilar membranes, which act as harmonic analyzers. (A broad introduction to the mathematics of musical sound can be found in the first eight chapters of David Benson's book [2].)

Although elements such as rhythm, melody, and harmony are frequently described as fundamental "dimensions" of music, the case can be made that in fact *timbre* (or *tone color*) is the most important universal musical quality: The strike of Pablo Casals's bow to a cello string can send chills up the spine, and Nat Cole's voice can convert a single syllable into the sublime. In this case, Fourier theory provides a strong mathematical explanation for this musical phenomenon: namely, that the timbre of a sound—which closely corresponds to the frequency spectrum of its wave shape—lives in an infinite-dimensional space! Well, the space is infinite-dimensional in principle, but even taking into account the limited frequency range of our conscious hearing (20 Hz–20,000 Hz), just a single second of reasonably digitized musical sound requires tens of thousands of coordinates because even the short-term time evolution of wave shape is critical to the perception of tone quality. The depth and complexity of timbre is further illustrated by the extreme difficulty of synthesizing musically interesting sounds by directly prescribing wave spectra and by the fact that a pure sine wave corresponds to a completely boring musical sound.

Of course, almost all musicians remain blissfully unaware of the elegance of Fourier theory as they coax out expressively complex sounds from traditional instruments, guided only by the analysis provided by their own ears. It is true that, by electronically synthesizing "unnatural" spectra, it is possible to generate sounds that cannot be made by traditional instruments—perhaps following a musical analogue of studying nonstandard axiomatic systems in mathematics—but such variations are not ends in themselves and have value only if they lead to "interesting" results.

Although timbre is fundamental to music, extending musico-mathematical relationships becomes problematic as sequences of sounds are extended in time and begin to acquire musical meaning. For instance, the well-studied relationships between whole number ratios and consonant pitch intervals, although interesting from a physical point of view and historically important, ultimately do not correspond to any cohesive *mathematical* notion, as musical aesthetics rightfully lead to compromises and approximations in choices of scales and tunings, with the resulting widely accepted equal-tempered chromatic scale, which has frequency ratios of $\sqrt[12]{2}$ for all pairs of adjacent notes. Explanations of this phenomenon are readily available, for instance in Chapter 5 of [2], as well as in Ian Stewart's delightful expository piece in Chapter 4 of [6], describing how a classical construction for placing the frets on a guitar ties together discussions of Pythagorean and equal-tempered scales, ruler and compass constructions, continued fractions, and fractional linear approximations of exponential functions. Although a small minority of musicians are obsessed with subtleties of tuning choices and justifications of scale constructions, the vast majority of musicians have no trouble making beautiful music with the equal-tempered pitch system, easily incorporating together instruments that have fixed tunings with those that are more flexible and happily exploiting the freedom to modulate between unrelated keys that is afforded by "theoretically compromised" scales. In any event, many instruments are tuned by hand, and notes are bent by ear, so it is not surprising that once musical flow commences, mathematical imperfections in pitch fade into the background.

Perhaps the irony that ancient hopes for combining rational numbers and music into a cohesive worldview have been dashed by the general acceptance of a musical system based on $\sqrt[12]{2}$ is an omen representative of problems that will haunt future attempts to build bridges between mathematics and music.

Analysis of Music

Three overlapping goals of music theory are to *explain why* music sounds the way it does, find good *ways to listen* to music, and describe *how to create* music. What might mathematics have to do with these goals? It certainly is natural to use permutations and transformations in describing *available* musical choices and relations between them (for instance, by representing pitch and rhythm in frequency–time coordinates or numbering scale tones relative to a root). But attempts at exhibiting substantial connections between *meaningful* musical choices and mathematics struggle to emerge from behind cloaks of terminology, perhaps precariously propped up by constructions of auxiliary geometric objects. The problem is that mathematical content comes in the form of proven statements about well-defined structures, and attempts at "explaining" musical phenomena usually involve structures that are not well defined, with conclusions justified by carefully chosen examples and multitudes of counterexamples ignored. And any logical development of well-defined structure is inevitably based on dubious or pedantic musical principles, so that the resulting conclusions can say precious little about what is important in music.

The types of problems illustrated in the basic examples considered here are compounded in more complicated analytic treatments of music.

MATHEMATICAL EXPLANATIONS OF MUSIC

For instance, the recent *Notices* articles [4, 5] use short-time Fourier transforms and continuous wavelet transforms to produce families of images from digital audio and claim to provide insight into musical structure that is both "quantitative" and "objective." The images do exhibit patterns that correspond to rhythmic accents, pitches, and volume, but the analysis of musical content is riddled with flaws and weaknesses that undermine most of the extremely enthusiastic conclusions.

The problems are well illustrated in Example 6 of [4], where four trivial musical observations are made about a short Duke Ellington excerpt: Sometimes symmetries appear in melodies, instruments can bend pitches, jazz can be syncopated, and melodies can contain varying groupings of notes. Areas of the associated images corresponding to these observations are located. It is claimed that "We can see from this analysis that this passage within just six seconds reveals a wealth of structure, including many features that are unique to jazz. Such mastery illustrates why Duke Ellington was one of the greatest composers of the twentieth century." The implication that the examination of the images illustrates anything about the music (let alone the greatness) of Duke Ellington is unfounded for several reasons.

First of all, the "analysis" admittedly includes listening to the recording; the note blobs in the image only contribute frequency readings from one coordinate and indicate rhythmic placement along the time coordinate. The observation of a slurring of pitch together with a brief descending–ascending motif leads the authors to conclude that

Ellington is synthesizing "a melodic characteristic of jazz (micro-tones) with one of classical music (reflection about a pitch level)." This conclusion, besides being musically trivial, ignores the fact that symmetries of melodic fragments and bending of pitches (not to mention syncopation) occur in all kinds of music—certainly in both jazz and classical music.

The fourth observation refers to a notion of "hierarchy" as giving "preferred" groupings of musical notes via grammar-like rules. But this notion of hierarchy is not well defined, as even recognized in [9] by the authors who coined the notion. And surely the "wealth of structure" visible in the images could also be created by a mediocre or even poor performance of the same or a similar piece. In fact, much richer visual structures could certainly be created by sounds that are more complicated, including sounds that are essentially devoid of musical content. No control examples are given, and the visual data require listening for interpretation, yet it is claimed "most importantly" that the images "provide an objective description of recorded performances." What does "objective" mean here? Are the authors suggesting that looking at their images provides some true measure of music? Even putting aside the trivial nature of the musical observations, this paragraph makes clear that any meaningful conclusions are in fact being *entirely drawn from listening.*

Example 6 of [5] implies that the images provide an answer to the question: What do Beethoven, Benny Goodman, and Jimi Hendrix have in common? The evidence of "approximate mirror symmetry" is only the trivial observation of melodic lines that descend and then ascend, a property of music that is probably familiar to even the untrained casual listener. Again, all kinds of sounds, including nonmusical ones, could give rise to similar images, and the restricted set of examples contained in [4] and [5] surely reflects the fact that extracting any meaningful general correspondence between the visible patterns and musical content is highly unlikely.

Acclaimed as the first musico-mathematical article to appear in *Science* magazine, [13] claims to illustrate how composers "exploit" the geometry of an orbifold and to show "precisely how harmony and counterpoint are related." Although this article contains well-defined statements and arguments, the weakness of the underlying musical principles erodes any meaningful connection with mathematics. The entire construction is based on the notion of "efficient voice leading," which is justified by the statement that "Western pedagogues instruct composers to minimize voice leading while eschewing crossing changes." In fact, this extremely limited notion can be considered relevant only when it is desired to have an accompaniment that is musically *benign* so as not to interfere with other concurrent statements and is at best a rule of thumb for a student composer and/or student arranger. The experienced creator of music certainly hears every voice and is guided by what sounds best rather than instructions from pedagogues. So, even ignoring some other questionable musical assumptions, it is difficult to derive any conclusions from a geometric construction that is based on a principle that "minimizes" musical content.

Other examples of "geometric" analyses are common, and musical scores written in the time and pitch coordinates of standard notation provide a plethora of patterns and data. The discovery of symmetries and other transformations of musical motifs (as notated) is often presented as evidence of an underlying mathematical component of music. But such discoveries do not correspond to musically coherent or mathematically interesting notions. While repetition and variation pervade music, precise symmetries among musical phrases are certainly not generic, so if such symmetry were musically meaningful, one would expect it to have a recognizable effect. But convincing counterevidence is provided by J. S. Bach's completely palindromic Crab Canon from his Musical Offering. What is remarkable about the Crab Canon is that even the most diligent listener is not going to have a clue that the piece is palindromic without access to the score, and in spite of the extreme notational symmetry, the piece sounds characteristically Bach-like and—by Bach's standards—less memorable than average. (In this case, Bach's compositional tour de force is in response to a challenge from Frederick the Great; more on composer-embedded musical patterns is discussed later.)

A method commonly employed in mathematical analyses of music (including [13]) is to identify pitches that differ by a whole number of octaves, and the resulting equivalence classes are assumed to be a natural object of study. Although it is true that pitches that are an octave apart have a clear notion of "sameness" (which is reflected in their shared overtones), the *musical* effect of changing the register of a note (choosing a representative of the pitch class) is not at all negligible. This notion suggests an interesting experiment: Listen to musical pieces whose pitch class representatives have been randomly permuted. Such shuffling of notes will certainly generate some bizarre-sounding music, and it is a safe bet that your favorite listening would lose its special place in your heart if always subjected to having its notes scattered in this way. But any musical theory that takes seriously the idea of working with pitch classes applies equally to "explain" such sounds! This modding out by octave "translations" is often invoked by music theorists to construct tori as parameter spaces.

MATHEMATICAL WAYS OF LISTENING TO MUSIC

The second goal of musical analysis raises an interesting question: How does extramusical information affect the listener? The effects are certainly wide ranging, from the relatively benign influences of knowing a song title or anecdotal stories about the performer to the enrapturement of an associated religious ritual. Lyric content or dance generally tends to interact strongly with accompanying musical statements, and when music is presented with video, the music likely plays a subservient role (and in such a setting the power of sound to generate its own images has been compromised). In the case of mathematically oriented music theory, it is usually tacitly assumed that an awareness of any "explanatory" mathematical notions improves the musical experience. Although this may be true for some music theorists, it is important to recognize that it is not necessarily a mathematical insight into essential general musical properties, but more likely a personal enhancement for one who enjoys attaching intellectual constructions to music. In fact, it can often be beneficial to remain ignorant of extramusical information, even when it is provided by the composer. More than once, I have been inspired by music accompanied by lyrics in a language I did not understand only to discover later that the words were not just unrelated to my appreciation but even unappealing to me. More generally, it is remarkable how in spite of the strong link between music and its ambient culture of origin, appreciation of music can bridge wide cultural gaps. For instance, secular appreciation of religious music abounds, the blues can go over well in Asia, hip-hop pieces are sometimes based on loops from classic jazz recordings, and "world music" has its own category in the commercial music market. The point here is that, although music comes wrapped in webs of extramusical connections, it is a subtle matter to extract *essential* threads from the midst of the many personal ones.

The effects of imposing conscious listening techniques often appear in the setting of music pedagogy: The journey from student to professional musician usually involves many years of music theory in the form of organizing sounds into recognizable bits and studying how they interact (there are many methods for doing this). This process of intellectualizing about music is often difficult because the student can become hypercritical and overly self-conscious, both as a performer and as a listener. Eventually the experienced musician is able to return to the appreciation of sound for its own sake, retaining the ability to analyze tension and resolution in theoretical terms at will but also free to enjoy the transcendental in-the-moment nature of music.

To clarify, I'm not proposing that analytic listening, mathematically motivated or otherwise, is wrong, just that it is not fundamental to the appreciation of music in general. All kinds of attentive, repeated, and earnest listening can access the full range and depth of musical meaning that is present in sound.

CREATION OF MUSIC

The most effective use of theory in the creation of music is to provide frameworks for "experimentation" rather than rules to be followed. Again, the methodical organization of sound may motivate the use of mathematical terminology, but although the resulting explorations may help the practicing musician gain insights into subtleties of musical tension and resolution, they are not going to lead to meaningful theorems expressing general essential musical qualities. In fact, even completely arbitrarily formulated methodologies can spark fruitful musical studies (and sometimes give birth to "styles" and "schools") merely by reducing the profusion of available musical choices.

For instance, the various serial composition techniques developed by Western atonal composers such as Schoenberg a century ago involve applications of various formal rules that were designed to avoid traditional combinations of sounds and can be described using elementary mathematical notions like transformations and permutations of pitches and rhythms. But this formalism expressed a self-conscious rebellion against tonality rather than any natural musical structure, and the value of the resulting music always depended, not surprisingly, on the creativity of the composer rather than (or in spite of) the formal structure. By mistaking rigidity (in the colloquial sense) for rigor (in the mathematical sense), such musical formalism is often presented as a "mathematical" aspect of music (e.g., Chapter 8 of [6]). The importance of twentieth-century formalist schools in music has been greatly exaggerated by academics, while the incorporation of dissonance and breaching of harmonic boundaries have proceeded more naturally in the rest of the vast musical world.

Although it is not surprising to the mathematician that arbitrary formalism is not mathematics, there is also music that has been created using constructions ostensibly based on mathematical elements (with varying levels of seriousness). However, the inevitable insertion of aesthetic choices, together with the arbitrary nature of the underlying constructions, conspires to remove any trace of mathematical content from the picture. For instance, examples of "fractal music" range from melodic fragments simply superimposed over themselves at a few increasing multiples of tempo to multiply-iterated computer synthesis of sound from two-dimensional fractal-like shapes that involves numerous parameter choices. The "poorer approximations" of fractals actually tend to sound more musical, but in any event results certainly do not inspire repeated listening and seem unlikely to produce anything nearly as interesting as properties such as fractional dimension, let alone to correspond to any more substantial fractal-related mathematics.

The relationships between the motivations and outputs of artists can be subtle and wide ranging. In the case of mathematically inspired composers, it's frequently a matter of "a little knowledge being a dangerous thing," and even for the mathematically astute creator of music there remains the problem of extracting correlation from the inspiration. For instance, when a composer claims that the Fibonacci sequence is essential to one piece of music and then turns around and embeds names into the next piece via rhythmic Morse code, the transient nature of any musico-mathematical relationships is apparent [3]. It is possible to be sincere without being serious, but it is also true that in some circles it can be advantageous for a musician to have a supporting "theory" that critics can latch on to.

Unfortunately, I've yet to hear any mathematically inspired music that comes close to providing the substance and lasting impression of even an elementary piece of reasonably interesting mathematics. This situation reflects a common occurrence in the art world, where the desire to innovate leads to the celebration of "newness for newness' sake," a phenomenon much less prevalent in mathematics, where the value of new work emerges by consensus rather than by press release and both the audience and the reviewers are mathematicians.

Metaphorical Comparisons

So if the physics of sound is mathematical but not musical and music theory is musical but not mathematical, we can still ask if a common musico-mathematical core is reflected in other, perhaps more metaphorical, ways. Attention is focused on the question of what might be special to mathematics and music rather than science and art in general.

Fundamental Observation

An interesting web of definitions, theorems, proofs, and conjectures does not require an extramathematical application to be satisfying. Similarly, the rhythmic flow of sonic tensions and resolutions in an instrumental music performance can be appreciated without attributing to the sounds any worldly connotations. In this respect, mathematics and music seem to share the property that their content—however subjective and time-dependent—*can* be expressed intrinsically, without direct reference to the natural world of human experience.

Whether you agree or disagree with this statement at face value, I believe it is worth trying to adjust your philosophical viewpoint enough to consider the claim, if only to clarify its limitations. (For instance, if you can't separate any significant part of mathematics or music from the natural world, then at least try to recognize the presence of a significant *degree* of intrinsic meaning.) Since I believe that this observation is important, some clarifications are in order.

First of all, there is clearly an emphasis on "can" because both mathematics and music frequently do refer directly to the natural world. While the mathematician is well aware of the subtle and symbiotic interactions between the abstract development of theories and applications of mathematics, analogous interactions also occur with music, which besides being appreciated for its own sake can be associated with lyrics, images, dance, ritual, ceremony, commerce, and other extra-musical phenomena. Of course external models are enriching and vital to both disciplines, but it can be helpful to be aware of the distinction, and I believe that the claimed observation of intrinsic meaning provides a special link between mathematics and music.

Among human disciplines, this form of intrinsic meaning is essentially *unique* to mathematics and nonlyric music: Other sciences are always directly tied to the natural world via their subject matter, and although other art forms may use abstraction, it almost always involves recognizable elements of human experience that have been distorted or used in unexpected ways.

It is true that certain visual art that is completely devoid of any reference to the natural world can have content, but I feel that the *general* comparison is not even close and that the intrinsic natures of music and mathematics are a significant order of magnitude stronger, although I do not know how to measure this. Some fans of extremely abstract visual art may disagree with me here, and admittedly this may be evidence of a "gray area" where meaning emerges self-referentially from patterns, visual or sonic, perhaps suggesting analogies with certain musical works that seem not to even reference recognizable elements of music. Also relevant here is that the visuals used by mathematicians to express mathematics, such as figures, graphs, and diagrams, can have an aesthetic effect of their own, as recognized for instance by the sculpture of Helaman Ferguson (http://www.helasculpt.com/). Some people might suggest that such images provide more effective artistic embodiments of mathematical ideas than the "pseudorigorous" mathematically inspired music composition techniques discussed here. In any event, I stand by the claim of a significant sense of uniqueness and continue with clarifications.

The locations, characters, and actions in literature and dramatic performance provide essential identifications with the natural world, as even the most fantastic settings inevitably mirror recognizable elements in the lives of the audience. And although the art lies in the development of tension and resolution through changes in relationships among the agents, the effect on the audience is always dependent on qualities and expectations that are inferred from these identifications.

And if the avid poetry listener feels that sometimes the message of the poem is being carried entirely by the cadence, phrasing, texture, and tone of voice of the poet without recognition of any semantic content in the words, then I'd say that what is being heard is *music*. Logical philosophy and computer science can similarly intersect mathematics at their extremes.

This claim of uniqueness is not a denial that other disciplines can have meaning that transcends their inherent references to the natural world but rather just the assertion that what is special to mathematics and music is that their content is capable of being expressed entirely in terms of their own raw material, namely, logical thought and audible sound.

Furthermore, no strict formalist mathematical philosophy is being imposed here, just the acceptance that the contemplation of generalized homology theories, transfinite ordinals, moduli spaces, and the like can (and often must) take place outside the usual realm of sensory perception. We believe that our elements are well defined, that our arguments are satisfyingly checkable, and that mathematics is consistent (although we know we can't prove it). Theories are developed by various internal associations of mathematical elements, but we do not *require* confirmation from an embodiment in human experience; and indeed we don't expect to find such confirmation, since even an object as basic as an interval of real numbers does not have a reliable model in the natural world.

Similarly, no banishment of cultural or other associations with music is being proposed, just the observation that as melodies, rhythms, and harmonies unfold in time, it is the relationships among the sounds that speak to you. The sounds repeat, mutate, diverge, return—always in combination with each other but never in *need* of "pointing" to anything outside the music.

Notice that such frequently recognized qualities as beauty, elegance, power, economy, anticipation, surprise, tension, and resolution are certainly *not* unique to music and mathematics. What is remarkable is that such qualities can emerge at all without need of body language, radiant sunsets, death-defying feats, wireless capabilities, expected rates of return, time travel, or love lost and renewed.

Finally, the claimed uniqueness of intrinsic meaning is not intended to imply any judgments on the relative *values* of human endeavors, any of which can of course have a wide range of appeal and utility to a variety of people. In particular, nothing is being implied about the relative importance of "pure" and "applied" in both mathematics and music.

WHAT DO METAPHORICAL OBSERVATIONS EXPLAIN?

The fundamental observation seems to provide a possible reason for the enduring attraction of musico-mathematical investigations: Since the ubiquity and power of mathematical and musical applications are a consequence of the strength of their intrinsic constructions, it is only natural to ask the question, "Can they model each other?"

But this very modeling power can represent obstructions to an in-depth metaphorical discussion with a general public whose musical and mathematical experiences are dominated by applications. (For instance, instrumental jazz and classical music each account for just a few percent of music sales, which is of course still greater than the publishing share of mathematics journals.) It is an important challenge to somehow share the value of abstract thinking with society at large.

An admirably well-intentioned attempt to describe metaphorical connections between the "inner lives" of music and mathematics to a general audience is the recently reprinted bestseller *Emblems of Mind* [12] by *New York Times* journalist Edward Rothstein. On the positive side, this book brings many worthwhile points to light, including the roles of beauty and creativity in mathematics, the emphasis of relationships over objects, and the power of abstraction inherent in both disciplines. Unfortunately, several fundamental problems cripple the coherent development of the many good ideas present. For instance, the occasionally insightful descriptions of music repeatedly fall into all the traps of musical analysis discussed above. A harbinger of the forthcoming distortion appears in the introduction, where after mentioning musical affinities of Galileo, Euclid, Euler, and Kepler, the author includes Schoenberg, Xenakis, and Cage among a short list of examples that seem to point back from music to mathematics. Even most mathematicians with an affinity for these composers would, with all due respect, surely recognize that this juxtaposition is way out of balance. This comparison leads to such contradictions as claiming the existence of "a systematic logic that guides musical systems" but then admitting later that great musical compositions "create their own form of necessity, the binding coming not from logic but from the unfolding of ideas . . ." And the spurious metaphorical equating of the contrived formalism of twentieth-century atonal "systems" with the discovery of non-Euclidean geometries both fails to recognize the strong and natural role of modern geometry in mathematics and sidesteps the truth that the natures of tonality and dissonance in music are complicated and mysteriously subtle phenomena that have defied satisfactory explanation by any general theory.

The confusion created by mistaking musical form for content is compounded by being interwoven with an informal poetic analysis of music, frequently laced with fancifully chosen mathematical terminology. While the appreciator of well-written romantic prose may enjoy the exposition, those looking for more substance may be disappointed because the attempt to nail down details makes the metaphors less robust rather than stronger. For instance, the notion that a "composition proceeds to 'prove' itself" or the claim of an analogue of "completeness" (of a logical system) in music are signs that the discussion is deteriorating. This deterioration is confirmed when one of the text's central points relates a metaphorical sense of "truth" in music to musical "style."

One fact clearly underscored by the book is that ordinary human language is much better at conveying mathematical ideas than musical ideas. Although the feeling that music is "telling a story" is often intensely felt by both listener and performer, there is no known well-defined "grammar" of music; and if a picture is worth a thousand words, then the relation between music and language must surely be exponential. On the other hand, mathematics has its set-theoretic foundations expressed in the formal languages of logic, and among mathematicians, informal conversation is the most common method of communicating mathematics. Does this situation suggest that music is in some sense *more* abstract than mathematics?

The popularity of [12] does confirm that there is a healthily curious audience among the general public. One would hope that such readers could be encouraged to pursue their investigation of mathematics in the growing number of expository sources written by mathematicians, such as the recent *Princeton Companion to Mathematics* [7] (although the brief section on mathematics and music in [7] gives too much weight to the type of superficial musical analysis criticized above).

CREATIVE PROCESSES

One might summarize the essence of a general metaphorical view by the statement that "mathematics and music are the science and art of analogy." Although it appears to be difficult to extract more precision from metaphors, I believe that, by focusing on mathematical and musical *creative processes*, useful conclusions can be drawn.

In fact, the process of creating or discovering mathematics is in many ways analogous to a small-group jazz performance. This notion is evident in the real-time exchange of ideas among collaborators, spontaneously alternating lead and accompaniment roles, guided by a thematic problem, developing material statement by statement, pursuing tangential ideas, adapting to mistakes, being ready for unexpected results, and never knowing for sure if the original goals will be achieved. I believe that this analogy with musical improvisation is stronger than any picture of the mathematician as the solitary composer (although the most vital composers do capture the spirit of improvisation in their works) because there is a sense in which the nonperforming composer can rework the landscape to "force his or her theorems to be true" (but not necessarily "interesting"), whereas the improvisor must face the unforgiving judgment of the moment while traveling without a seatbelt. The analogy also extends to the researcher working alone as a solo improvisor, simultaneously playing lead and accompaniment roles as the devil's advocate, and even to the processes of understanding mathematics and interpreting composed music. (Note that the tradition of improvisation in Western classical music, which stretches back through Beethoven, Mozart, and Bach, shows signs of a rebirth [11].)

But this improvisational analogy can apply more generally to processes involved in many human endeavors, not only in the arts and sciences but also including many workplace environments encountered by citizens of today's fast-changing global society. In fact, in the face of turbulent economic conditions, advancing technologies, and increasingly international markets, employers and employees alike are going to be dealing with shifting work flows and new job types and products, as well as interactions with foreign cultures, all of which require creative problem solving to recognize appropriate skill sets, implement effective training and study methods, and develop new career and employment programs.

The key point here is that the intrinsic nature of mathematics and music alike suggests that the studies of both research mathematics and improvisational music could play valuable roles in modern education, as their *abstract yet cohesive structures serve as models* for developing flexible skills and the ability to generate spontaneous constructive thought.

Whereas the problem-solving techniques and computational powers of mathematics are already well appreciated, the more abstract, creative, and improvisational aspects of human thought are going to be increasingly valuable in twenty-first-century life.

Ideally these studies would be completely integrated into the education system, with the associated musical and mathematical learning processes naturally complementing and reinforcing each other. What is important here is that the *goals of research and improvisation guide the pedagogy*. The challenge is to develop courses, programs, and teaching conceptions with these goals in mind and to incorporate them into the curriculum. (Note that combining mathematics and music in the classroom is not being proposed here.)

That the underlying frameworks of the studies can complement and reinforce each other is apparent at many levels. For instance, the student of musical improvisation uses formalism (music theory) to generate examples (sounds) that are examined aesthetically (by listening), while the student of mathematics generates examples (special cases) to understand formalism (general statements) that are considered logically (by proving or disproving). More generally, both studies develop experience with solitary practice, group work, and open-ended learning. Many other such pedagogical frameworks exist at all levels and age groups.

The idea is not to produce more professional mathematicians and musicians (although talent would be more likely to flourish in this environment) but rather to provide greater general access and exposure to the relevant abstract skills. Of course, some students will benefit more from musical study, others from mathematics, and both subjects will still be challenging for almost everyone. But the recognition of the long-term benefits should provide motivation, and effective integration into the education structure would provide support to maximize the positive value for as many as possible. *The almost complete ignorance of the essences of mathematical research and improvisational music that is prevalent in society today means that the initial marginal benefits could be enormous.*

Of course, the challenges faced in implementing such an educational vision would be huge because effective teaching of both research mathematics and improvisational music is already difficult enough, and the skeptic may point to the existing body of inconclusive studies regarding musical and mathematical pedagogical methodology, as well as the apparent lack of supporting circumstantial evidence (where are all the improvisational music groups of mathematical researchers?). But there are good reasons to believe that the obstacles are surmountable and that the vision is valid. First of all, there is a growing consensus supporting educational reform, as well as funding available for innovative ideas. The mathematical research community has shown purposeful commitment to teaching in recent years, while at the same time the many jazz departments in universities and colleges across the country have become increasingly populated with top-level faculty who have significant performance experience. (So a pilot program for preparing teachers could involve cross-training of graduate and/or undergraduate students, for example.) And although music has been largely cut from primary and secondary school curricula, the many independent organizations that have been providing music instruction could provide infrastructure for pilot programs on the musical side. On the mathematical side, a new vision is desperately needed to guide a complete reforming of the current generally dreadful state of mathematics education at the primary and secondary school levels. That aspects of this vision have already been accepted is evidenced by the increasing numbers of mathematics Ph.D.s working outside academia [10] and by the direct implementation of jazz concepts in high-level business consulting [8].

Existing educational data should not be expected to provide insight into the worth of the proposed vision, primarily since such a focus on research and improvisation has not been significantly implemented. I would also expect that direct effects will be difficult to measure, especially in the short term. The problem of correlating success in varying job types is in itself an interesting problem in today's ocean of information and shifting employment patterns. And although I know of various successful external applications of musical and mathematical frames of mind, the satisfying nature of improvisational and research experiences means that those who are good at it are likely to happily stay with it.

Conclusion

It is clear that an in-depth appreciation of both mathematics and music is a prerequisite to the critical consideration of musico-mathematical relationships and their kernels. But to the extent that one appreciates mathematics, one *is* a mathematician, whereas the appreciator of music need not be a musician. It follows that the mathematics community is likely to provide constructive contributors to the dialogue. Expositing and teaching mathematics (independently of music), as well as promoting exposure to all forms of music, will contribute to opening the discussion to a wider audience. Ideally this teaching could be integrated into the entire educational system. At least, one would hope that inviting metaphors might provide motivation for deeper exploration and in particular lead to a wider awareness of the aesthetics of mathematics. The danger is that the unconscious readiness with which the mind accepts analogies will allow poetic hand waving to stir up pleasing but shallow illusions, clouding a picture that can only be clarified by thoughtful hard work.

In an ideal world, a marriage of mathematics and music should celebrate the beauty and power of abstraction. But the courtship is thrown off balance by the contrast between the open-access nature of the musical world, in which the listener is free to navigate by ear, and the rigor of the mathematical world, in which the curious mind must temper its imagination with logic. The proliferation of suitors in the natural world further complicates matters, rendering detailed agreements, scientific or metaphorical, elusive. In spite of the voluminous literature inspired by this undeniably intriguing situation, many of the most salient observations on the subject are one-liners, often provided by mathematicians (e.g., [1]). On that note, I would like to provide an affirmative answer to the title question by offering a punch line of my own: "Mathematics is like music that only musicians can hear."

Acknowledgments

The ideas presented here have been shaped by many conversations with people from a wide variety of mathematical and musical backgrounds. Particular thanks go to Paul Kirk (Indiana University) and Alex Barnett (Dartmouth College) for recent stimulating discussions.

References

[1] R. C. Archibald, Mathematicians and music, *The American Mathematical Monthly* **31** (1924), no. 1, 1–25. http://www.jstor.org/stable/2298868

[2] D. J. Benson, *Music: A Mathematical Offering*, Cambridge University Press, 2007.

[3] J. Bohannon, Riffs on numerical themes, *Science* 315 (2007), no. 5811, 462–463. http://www.sciencemag.org

[4] X. Cheng, J. V. Hart, and J. S. Walker, Time frequency analysis of musical rhythm, *Notices of the AMS* **56** (2009), no. 3, 344–360.

[5] G. W. Don, K. K. Muir, G. B. Volk, and J. S. Walker, Music: Broken symmetry, geometry, and complexity, *Notices of the AMS* **57** (2010), no. 1, 30–49.

[6] J. Fauvel, R. Flood, and R. Wilson (eds.), *Music and Mathematics*, Oxford University Press, 2003.

[7] T. Gowers (ed.), *The Princeton Companion to Mathematics*, Princeton University Press, 2008.

[8] J. Kao, *Jamming: The Art and Discipline of Business Creativity*, HarperCollins, 1996.

[9] F. Lerdahl, The genesis and architecture of the GTTM project, *Music Perception* **26** (2009), issue 3, 187–194.

[10] P. Phipps, J. Maxwell, and C. Rose, 2009 Annual survey of the mathematical sciences (preliminary report), *Notices of the AMS* **57** (2010), no. 2, 250–258.

[11] A. Ross, Taking liberties, *New Yorker Magazine*, August 31, 2009. http://www.new yorker.com/arts/critics/musical/2009/08/31./09o831crmu_music_ross

[12] E. Rothstein, *Emblems of Mind: The Inner Life of Music and Mathematics*, Times Books (1995), University of Chicago Press, 2006.

[13] D. Tymoczko, The geometry of musical chords, *Science* 313 (2006), 72–74.

Flat-Unfoldability and Woven Origami Tessellations

Robert J. Lang

The field of *origami tessellations* has seen explosive growth over the past 20 years. Interpreted broadly, an "origami tessellation" is a figure folded from a single sheet of paper in which the surface is divided up (tessellated) into a highly geometric pattern that is created by the folded edges and/or the transmission image of the varying layers (if folded from translucent paper and backlit), so that the pattern of the folded edges, rather than the outline of the figure, provides the dominant aesthetic. Though many origami tessellations are derived from regular tilings of the plane, the field of such 2D and 3D patterns is large and diverse. Similarly diverse are the algorithms by which they may be constructed.

For flat origami tessellations, one of the key constraints in their design is the *Kawasaki–Justin condition* (KJC), a condition found and described independently by Toshikazu Kawasaki [Takahama and Kasahara 85] and Jacques Justin [Justin 86]. There are several equivalent formulations, but the most common is the following: A crease pattern can be folded flat only if, at every interior vertex,

$$\alpha_1 - \alpha_2 + \alpha_3 - \alpha_4 \ldots = 0, \tag{1}$$

where the $\{\alpha_i\}$ are the sector angles around the vertex, numbered cyclically. The Kawasaki–Justin condition is not *sufficient* for flat-foldability; additional conditions apply to the crease assignment and layer ordering (also formulated by Justin [97]). However, in many situations, the primary challenge in designing an origami crease pattern is ensuring that it satisfies KJC.

The design of an origami figure, be it a tessellation or a representation, is typically framed as an *inverse problem*: The desired folded form or some aspect(s) of it are specified, and then the crease pattern that produces that folded form is constructed.

For a flat origami figure, if we know the folded form in its entirety, then the design problem is essentially done: We can simply unfold the folded form, either algorithmically or physically, using real paper. But in the design process, we rarely know the complete folded form at inception. Rather, we specify some elements of the folded form based on design criteria and then choose other elements of the folded form to ensure *validity*. Validity, in this case, means that the folded form comes from a flat sheet of paper, as opposed to some unusual, nonflat shape.

KJC applies to a *crease pattern*; it ensures that a crease pattern results in a flat form when it is folded. To solve an inverse problem, however, we want to find a condition on a *folded form*: the conditions on its mathematical description that ensure that it *unfolds* to a flat sheet of paper. Such a condition is called a *flat-unfoldability* condition. This paper introduces and describes a broadly useful flat-unfoldability condition.

The utility of a flat-unfoldability condition is this: If we can design a folded form that achieves a particular design goal and that satisfies flat-unfoldability conditions, then we can simply mathematically unfold the folded form to realize the crease pattern that we need to start with. We demonstrate this approach with the solution to a previously vexing origami design problem: the design of a general woven origami tessellation.

Woven Tessellations

One of the simplest origami tessellations is the alternating *simple flat twist tessellation,* composed of twisted squares in which adjacent squares rotate in opposite directions. In such a tessellation, it is possible to assign creases so that all of the squares are surrounded by mountain folds (i.e., they are all in cyclic form). An example of such a tessellation is shown in Figure 1.

If you invert the crease assignment in such a tessellation (or turn it over), a surprise is in store: the pattern of edges looks uncannily like a set of woven strips, as shown in Figure 2. This is a pleasing illusion because origami tessellations are, by definition, folded from a single sheet, but this structure looks like it was constructed from many separate strips of paper (plus a background field).

Figure 1. Crease pattern for alternating simple flat twist tessellation consisting of square twists with cyclic mountain crease assignment (left). The folded form of this tessellation with translucent paper (right).

Figure 2. Crease pattern for an alternating simple flat twist tessellation consisting of square twists with cyclic valley crease assignment (left). The folded form of the tessellation with nearly opaque paper (right). This crease assignment displays the appearance of continuous woven strips.

This particular pattern suggests that there might be a family of sub-patterns as we start to consider generalizations (several people have done so; Bateman [10a and 10b] show some examples). In the square woven tessellation, the strips run up and down and side to side and are evenly spaced. But one could envision patterns in which the strips are at other angles, or other spacings—or even at no particular angle, with every strip running in a different direction. In addition, if we think of such patterns as "patterns formed by woven strips," we not only have the spacing and angles of the strips to choose from: we can also choose, at every intersection, which strip appears to be on top.

We can also choose the width of the apparent strips, and we can, in principle, choose them all independently. These variables outline a vast space of potential patterns, any of which may or may not be realizable as a single-sheet origami tessellation. Going forward, I call such patterns *woven origami tessellations.*

The descriptions of the patterns possible with woven origami tessellations parallel the patterns possible in textile weaving: in both cases, one can vary the strips (analogous to the warp and weft threads in textiles), their widths, angles, and the pattern of crossings, or how the various strips and threads cross over one another. The simplest woven pattern is called a *plain weave*, or a *simple over-and-under* weave, in which any given strip, followed along its length, alternately goes over and under the strips that it crosses. If we further stipulate that

- no more than two strips cross at a given point,
- every strip travels in a straight line,

then we can narrow the field of possible woven tessellations considerably. I call any tessellation that uses a simple over-and-under weave and that satisfies these two conditions a *simple woven tessellation.*

The question then arises: What are the possible simple woven tessellations, and how are they folded using origami?

Simple Woven Patterns

To further simplify matters (and to make an aesthetic choice), let us assume that all strips are the same width. Then we can construct a simple woven pattern by the following prescription, illustrated in Figure 3.

1. Choose some pattern of straight lines such that no more than two intersect at any vertex.
2. Fatten each line to the desired width of the strips.
3. Add a boundary to the pattern to define the background field.
4. At each crossing, erase two of the four lines at the crossing to create a simple over-and-under weave.

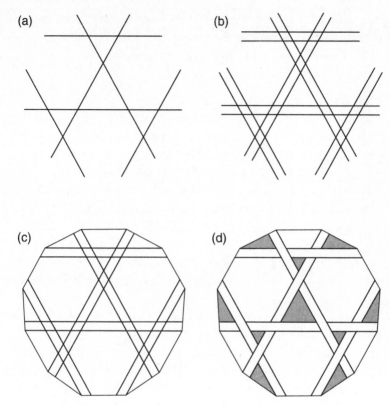

Figure 3. Construction sequence for a simple weave: (a) begin with a pattern of lines for which no more than two intersect at any given point; (b) thicken each line; (c) add a border to the field; (d) selectively erase crossings based on a two-coloring of the polygons between the woven edges.

How to do this last step is, perhaps, not entirely obvious, but a simple procedure suggests itself. Note that all interior vertices of the line pattern have degree 4, which means that the pattern can be two-colored, as shown in Figure 3. Each polygon is surrounded on all interior sides by partial strips. If we give each polygon a counterclockwise (CCW) circulation, each side of the polygon has a *head*, which is the end of the strip segment in the CCW direction, and a *tail*, which is the end of the strip segment at the other end. We can use these definitions, plus the two-coloring, to create the over-and-under woven pattern as follows:

- In colored polygons, the head of one strip segment goes under the tail of the next strip segment.
- In white polygons, the head of one strip segment goes over the tail of the next.

How do we turn the woven pattern into an origami figure? We can get an idea of this by looking at the square pattern again, but this time let's give the paper some translucency so that we can see the hidden layers of paper—which is where all the action is—and we'll zoom in a bit. Figure 4 shows the crease pattern and folded form.

Pay particular attention to the highlighted region in Figure 4. We see that this shaded trapezoid reappears throughout the pattern; in fact, every crossing of two strips has one of these trapezoids. Every trapezoid has two obtuse-angle vertices, which are the (hidden) ends of a going-under strip, and two acute-angle vertices, which are the endpoints of a covering-up strip. Then, of course, each of the acute-angle vertices of a trapezoid is incident to an obtuse-angle vertex of an adjacent trapezoid. When the pattern is unfolded, each trapezoid appears explicitly in the crease pattern.

In principle, we could imagine that any woven tessellation pattern might be realized by placing some version of this same trapezoidal structure at every strip crossing in a woven tessellation. Using our test pattern of strips from Figure 3, a hypothetical example is shown in Figure 5.

The next question is: Can this folded form pattern be folded from a flat sheet of paper?

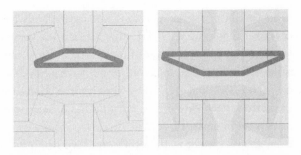

Figure 4. Crease pattern (left) and folded form (right) with the key trapezoidal structure highlighted.

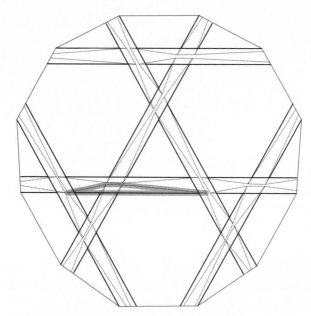

Figure 5. A hypothetical folded form of the woven-strip pattern from Figure 3.

Flat-Unfoldability

The question just posed is a specific example of a more general problem: If we are given a description of a flat-folded form (complete or partial), what are the conditions that ensure that it can be folded from a simple flat sheet of paper? This question is an inverse problem, and its answer would be the opposite of the more well-known flat-foldability conditions that determine whether a crease pattern can be folded into a flat-folded form. Those conditions are KJC (that the alternating sum of the angles at any interior vertex equals zero) and Justin's layer-ordering conditions (that disallow self-intersection). KJC is a *metric* condition; it ensures that the folded form lies flat when constructed from non-stretchy (isometric) material. Justin's conditions are combinatorial, and because they describe the layer ordering of the folded form, they are, implicitly, a condition of the folded form.

If a folded form satisfies Justin's conditions—we have chosen a layer ordering that avoids self-intersection—then the only condition still to be met is some isometric condition, and the particular isometric condition

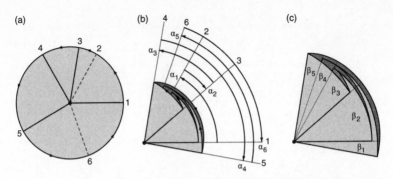

Figure 6. (a) Crease pattern for a vertex. (b) The measured angles of the folded form from one vertex to the next. CCW angles are positive; CW angles are negative. (c) Angle β_i is the angular extent from the ith fold to the next.

would be that the sector angles around each interior vertex, when the paper is unfolded, must sum to 360° so that the vertex lies flat.

Given a folded vertex and the information about each layer, we can define the sector angles of the vertex, in order, as the rotational angle from each folded edge to the next within each separate layer. If we adopt the usual convention that CCW rotation is a positive angle, then we find that successive sector angles in the folded form alternate in sign: first positive (CCW), then negative (clockwise, or CW), then positive, and so forth, until we reach the folded edge at which we started. If we label these *folded* sector angles $\alpha_1, \alpha_2, \ldots$, as shown in Figure 6(b), then the condition that we end up where we started is simple:

$$\alpha_1 + \alpha_2 + \alpha_3 + \alpha_4 \ldots = 0. \tag{2}$$

But if this folded vertex arose from a flat sheet of paper, when we unfold the vertex, we should get 360 degrees of angle around the unfolded vertex, and so the condition that the folded form sector angles must satisfy is

$$\alpha_1 - \alpha_2 + \alpha_3 - \alpha_4 \ldots = \pm 360°, \tag{3}$$

where the sign of the result depends on whether we started with a positive or negative angle. This equation is the *flat-unfoldability condition*, analogous to KJC for flat-*foldability*.

To apply Eq. (3), one must be able to identify the layers incident to each folded edge to construct the cyclic ordering of the sector angles.

However, sorting the order of the folded edges is not necessary. If we examine the folds that emanate from the vertex, they always lie strictly within a 180° arc. Starting at one end, we take β_i to be the angular extent of the ith arc, and we denote by n_i the number of layers of paper *that are incident to the vertex* (i.e., we don't count layers that are not part of the vertex figure). Then, since every layer incident to the vertex must form part of the flat, unfolded vertex, it must be the case that

$$\sum_i n_i \beta_i = 360°. \tag{4}$$

For the commonly encountered degree-4 vertex, the pattern of fold lines falls into one of three possible patterns, and it is possible to construct special cases of Eq. (4) that apply to the line pattern of the folded form. Every flat-foldable degree-4 vertex falls into one of the following cases (Figure 7):

1. all four sector angles are distinct,
2. the sector angles come in two pairs of equal angles,
3. all four sector angles are equal to 90°.

If all four sector angles are distinct, then there are four distinct lines in the line pattern of the folded form, and the largest angle in the line pattern less than 180° is one of the four sector angles, as shown in Figure 7(a). If we number the three visible angles within this angle by $\{\beta_1, \beta_2, \beta_3\}$, then the four sector angles in the crease pattern are, respectively,

$$(\beta_2), (\beta_1 + \beta_2), (\beta_2 + \beta_3), \text{ and } (\beta_1 + \beta_2 + \beta_3), \tag{5}$$

and so the flat-unfoldability condition applicable to the visible angles in the folded form is

$$2\beta_1 + 4\beta_2 + 2\beta_3 = 360°, \tag{6}$$

or, equivalently,

$$\beta_1 + 2\beta_2 + \beta_3 = 180°. \tag{7}$$

If the sector angles come in two pairs of equal angles, then there are three distinct lines in the line pattern of the folded form, and the largest angle in the line pattern less than 180° is equal to two of the four sector angles, as shown in Figure 7(b). Numbering the two visible angles within this angle by $\{\beta_1, \beta_2\}$, the four sector angles in the crease pattern are, respectively,

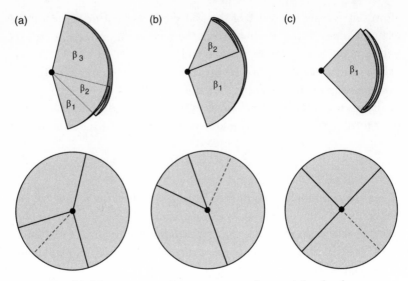

Figure 7. Folded form (top) and crease pattern (bottom) for the three distinct configurations of a degree-4 vertex: (a) all four sector angles are distinct, (b) sector angles form two pairs of equal angles, (c) all four sector angles are equal to 90°.

$$(\beta_1 + \beta_2), (\beta_1 + \beta_2), (\beta_2), \text{ and } (\beta_2), \tag{8}$$

and so the flat-unfoldability condition applicable to the visible angles in the folded form is

$$2\beta_1 + 4\beta_2 = 360°. \tag{9}$$

And finally, if all four sector angles are equal, there are two distinct lines in the line pattern of the folded form; the angle in this pattern less than 180° is equal to each of the sector angles of the crease pattern, as shown in Figure 7(c), and so the relationship between the sector angle of the crease pattern and the corresponding condition on the visible angle $\{\beta_1\}$ of the folded form will be

$$\alpha_1 = \alpha_2 = \alpha_3 = \alpha_4 = \beta_1 = 90°. \tag{10}$$

Thus, given the line pattern of a folded form composed of degree-4 vertices and a valid layer ordering, the folded form can be unfolded to a flat sheet of paper if and only if for every interior vertex of the line pattern, one of Eqs. (6)–(10) (as appropriate) is satisfied (and there is no self intersection). There is no guarantee, of course, that any such pattern can be rigidly unfolded.

Parameterizing the Woven Tessellation

To satisfy the flat-unfoldability conditions, there must, of course, be some variables whose values can be adjusted to satisfy the equations. (One hopes that there are enough adjustable variables to satisfy *all* of the equations that must be satisfied.) In the woven tessellation, at each strip crossing, there are two trapezoids, each of which has four vertices. Each vertex, however, appears in two different trapezoids (as an obtuse vertex of one trapezoid and an acute vertex of the other). Thus, if N_C is the number of strip crossings, then there are

$$\frac{1}{2} \times 4 \times N_C = 2N_C$$

flat-unfoldability conditions to be satisfied. That situation suggests that we should have at least that many variables in a parameterization of the crease pattern.

As it turns out, there is not total freedom in parameterizing the crease pattern because the base line (and thus the two acute vertices) of each trapezoid is required to lie on the "over" edge of a woven strip at a crossing and each of the obtuse vertices must lie on an "under" edge at a crossing. For each vertex, there is only one free parameter, which we can take to be, for example, the perpendicular distance of that vertex from the edge covering it. We call this distance the *inset distance d_i* for the vertex, as illustrated in Figure 8. If we do the same counting of variables, we find that the number of variables associated with each full trapezoid is also $2N_C$. So there are exactly as many variables arising from interior vertices as we have flat-unfoldability conditions.

In fact, though, there are somewhat more variables. Along the boundary of the folding pattern, we have partial trapezoids that are defined with more inset distances, but there is no need to satisfy a flat-unfoldability equality for vertices on the boundary of the folded pattern. (There is a requirement that the total angle be *less than or equal* to 360°, but that will not be an issue here.) So, in addition to the $2N_C$ variables associated with each of the crossings, we have $2N_B$ extra variables to work with, where N_B is the number of strips that hit the boundary in the folded form.

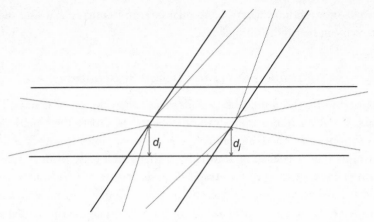

Figure 8. Schematic of a crossing vertex. d_i is the inset distance at each vertex.

The problem, then, is underconstrained; there are more variables than there are equations, and it is necessary to choose some additional criteria that allow one to solve for a particular solution. One could theoretically identify exactly the number of equalities needed to solve for all of the variables, but identifying the proper set can be a challenge. A more robust approach is to add one or more equality and/or inequality conditions that address any additional (perhaps aesthetic) criteria, then perform a multidimensional optimization with a suitable figure of merit to "soak up" any remaining degrees of freedom.

One additional criterion that would be useful is to set a minimum value on the inset distance; this criterion prevents the creation of trapezoids that are too skinny to be easily folded. Such a requirement takes the form of a set of inequality constraints:

$$d_i \geq d_{\min} \text{ for all } i. \tag{11}$$

Once a minimum inset distance is set, then we would also prefer that each of the trapezoids not be much *wider* than the minimum size in order to minimize the chances that two trapezoids overlap in such a way as to violate a self-intersection condition. This restriction can be accomplished by introducing a slack variable, $d_{\max}$, setting inequality constraints

$$d_i \leq d_{\max} \text{ for all } i, \tag{12}$$

and then taking $d_{\max}$ as the figure of merit to be minimized.

Another approach that has a similar effect of keeping the pleat size down and has the benefit of applying force to all vertices is to take the figure of merit to be the root-mean-square (RMS) sum of all inset distances and minimize that. In practice, I have found that this merit function works well and results in aesthetically pleasing patterns.

Conclusion

I have implemented the algorithm described above using Mathematica™ 7.0.1. The algorithm takes as input a pattern of lines that defines a woven pattern and a desired strip width (which must be small enough that there are no points where three or more woven strips overlap; auxiliary functions in the Mathematica™ notebook let the user solve for the maximum strip width for a given pattern of lines). The program finds all intersections between pairs of lines to turn the line pattern (plus a specified boundary curve) into a plane graph; it then two-colors the plane graph to determine the over-and-under pattern of the strips. At each strip crossing, it constructs the two trapezoids, suitably parameterized on the inset distances $\{d_i\}$. The user specifies a minimum inset distance (commonly half of the strip width), and then the inset distances are solved for, using the RMS-minimization optimization algorithm. The result is a folded form satisfying the flat-unfoldability conditions, which therefore can be unfolded to a flat pattern. Another Mathematica™ function takes the embedded graph of the folded form and algorithmically unfolds it to realize the crease pattern that gives rise to the desired folded form.

Figure 9 shows an example of this series, including the computed crease pattern, the computed folded form, and a folded example. Both the folded form and crease pattern in Figure 9 are computed, so there is no guarantee (beyond the mathematical arguments given above) that these two patterns really go together. However, this is, in fact the case; Figure 9 shows a photograph of a model folded from the crease pattern.

As a side note, these patterns tend to be very difficult to fold, and even though the starting state and ending states are both flat, the partially folded intermediate state is typically highly convoluted with the facets bent and/or even somewhat crumpled (one attempts to minimize the crumpling, of course). It is unlikely that one could find and successfully

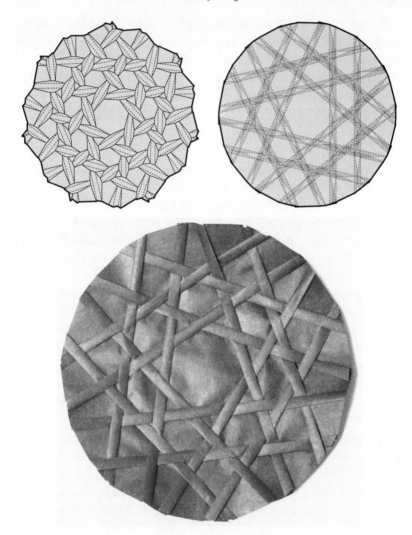

Figure 9. A woven pattern with sevenfold symmetry: computed crease pattern (top left); computed folded form, with translucent rendering to show the trapezoids (top right); a folded example (bottom).

fold even a moderately complex woven pattern by trial and error, but with this algorithm, any simple woven tessellation is now possible.

Since developing this algorithm, I have designed and folded a variety of woven tessellations of this family; a representative sampling may be found at Lang [10]. One can also envision various generalizations and

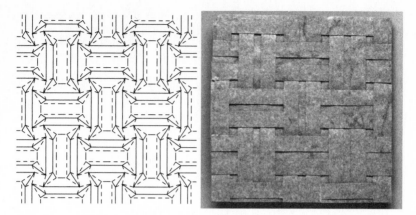

Figure 10. Crease pattern for a double-strip woven tessellation (left); photograph of a folded example (right).

variations; for example, instead of using an unconstrained boundary, one could implement periodic boundary conditions to realize a woven tessellation pattern that tiles the plane. But even beyond woven tessellations, I feel that the technique used here of applying flat-unfoldability conditions to a partially defined folded form pattern is a broadly useful tool for the design of geometric origami figures such as origami tessellations.

I close with a somewhat more complicated woven pattern, a "double-weave" pattern that appears to be composed of pairs of woven strips, shown in Figure 10. This pattern is more complicated than the simple woven pattern: there are both degree-4 and degree-6 vertices in the line pattern of the folded form (and therefore, of course, in the crease pattern). Nevertheless, it, and others like it, can be constructed in the same way, by applying flat-unfoldability conditions to a partially defined desired crease pattern. I expect that many more origami designs may be realized using this technique.

Bibliography

[Bateman 10a] Alex Bateman. "Not a Rattan Weave Tessellation." *Flickr.com*. Available at http://www.flickr.com/photos/42673512@N00/1513700768/, 2010.

[Bateman 10b] Alex Bateman. "Thin Origami Rattan Weave." *Flickr.com*. Available at http://www.flickr.com/photos/42673512@N00/1611849065/, 2010.

[Justin 86] Jacques Justin. "Mathematics of Origami, Part 9." *British Origami* 118 (1986), 28–30.

[Justin 97] Jacques Justin. "Towards a Mathematical Theory of Origami." In *Origami Science and Art: Proceedings of the Second International Meeting of Origami Science and Scientific Origami,* edited by K. Miura, pp. 15–30. Shiga, Japan: Seian University of Art and Design, 1997.

[Lang 10] Robert J. Lang. "Gallery (of Origami Tessellations)." *Flickr.com.* Available at http://www.flickr.com/photos/langorigami/, 2010.

[Takahama and Kasahara 85] Toshie Takahama and Kunihiko Kasahara. *Top Origami.* Tokyo: Sanrio Publications, 1985.

A Continuous Path from High School Calculus to University Analysis

Timothy Gowers

If I was asked to name the two most notable ways in which university-level mathematics differs from high school-level mathematics, then I would say that they were *abstraction* and *rigor*. Early courses at university in subjects such as group theory and linear algebra introduce students to the axiomatic way of thinking, while a first course in mathematical analysis introduces them to rigorous proofs of statements that they will hitherto have justified only informally, if at all. It is often claimed that mathematical analysis is difficult to learn because to understand it one must learn to think in a new way. In this short presentation, I would like to suggest that there are many connections between the advanced, rigorous way of thinking and the more naive way of thinking that would come naturally to a schoolchild. How these observations should influence the way we teach analysis is far from clear, but it cannot do any harm to draw attention to them.

I plan to discuss three aspects of basic real analysis: the axiomatic approach to the real number system, the definition of continuity, and the proof of the intermediate value theorem. In each case, I shall compare how they are treated in a typical analysis course (or textbook) with how they are thought of by an intelligent mathematician who has not yet attended such a course.

First, the real number system. The advanced attitude to the real numbers is this: There exists a complete ordered field; complete ordered fields can be constructed in many different ways; the mere fact that they exist is more important than the precise details of the constructions since any two complete ordered fields are isomorphic; therefore, it is best to treat the real numbers axiomatically, deducing everything from the axioms for a complete ordered field.

In practice, the fact that the real numbers form an ordered field is kept firmly in the background. We just add them, multiply them, take reciprocals of nonzero numbers, put them in order, and take for granted that they obey the obvious rules. In that respect, a university-level mathematician ends up behaving in a similar way to a high school-level mathematician, who also takes these various rules for granted (the difference is that a high school-level mathematician may well not have consciously thought about them).

What really separates the university mathematician from the high school mathematician is the use of the completeness axiom (in one of its forms). Or does it? What does the high school mathematician use instead? Does the high school mathematician even need a substitute, or is the completeness axiom just used for "advanced" statements?

Let us think about a few statements that need the completeness axiom in their proofs. One is the Archimedean axiom, in the form $n^{-1} \to 0$. To prove this, we say that the sequence is monotone decreasing and bounded below by 0. It therefore converges to a limit L, and a simple argument shows that L has to be 0.

An obvious difficulty for the high school mathematician is that the definition of convergence is not part of the high school curriculum. But the following equivalent statement is readily comprehensible at the high school level: For every positive real number x, you can find a positive integer n such that n^{-1} is less than x.

Now this last statement comes into the unfortunate category of statements that need a proof but that appear to the nonexpert to be bafflingly simple. Surely, a high school mathematician might say, all you have to do is choose enough 0s so that the number

$$0.000 \ldots 0001$$

starts with more 0s than x does, and then take n to be the reciprocal of this number. Or, even simpler, take the reciprocal of x and let n be the next integer above it. The university mathematician might then reply, "Ah, but you are assuming that every real number has a decimal expansion," or, "How do you know that there is *any* integer above it?" To which the high school mathematician will reply that a real number just *is* an infinite decimal (give or take pedantic qualifications about recurring nines) and that there is obviously an integer above it because you can just get rid of the fractional part of x and then add 1.

It is clear from these responses that the high school mathematician is thinking in terms of a model of the real numbers—defined in terms of infinite decimal expansions—rather than axiomatically. So perhaps there is a profound difference after all.

Before we accept this conclusion, let us think about another statement that a high school mathematician finds obvious: that there exists a positive real number x such that $x^2 = 2$. Why is this obvious? I think the (unarticulated) reason is this: They know in principle how to calculate it. They know that it is roughly 1.414, and they know that the reason for that is that 1.414^2 is a tiny bit smaller than 2, while 1.415^2 is a tiny bit bigger than 2. And the next digit is 2 because 1.4142^2 is an even tinier bit smaller than 2, and 1.4143^2 is an even tinier bit bigger than 2. And so on. (Moreover, each new digit can be found by a simple process of trial and error.)

There are a few hidden assumptions here, of course, most notably the continuity of the function $f(x) = x^2$. However, a high school mathematician is not too far wrong to find it obvious that the difference between 1.4142^2 and 1.4143^2 is very small and that as you add more and more digits, the corresponding differences get smaller and smaller. And if one imagines this process going on forever and producing a number with infinitely many digits, then what one is doing is not different from a rigorous proof by repeated bisection, except that in this case we do not really need an axiom to see that the monotone sequence 1, 1.4, 1.41, 1.414, 1.4142, . . . converges: It converges to the infinite decimal that has these finite decimals as its initial segments.

The way that a high school mathematician finds the decimal expansion of $\sqrt{2}$ can easily be converted into a proof that every real number has a decimal expansion. Of course, the resulting proof assumes the Archimedean axiom, so we cannot use decimal expansions to prove the Archimedean axiom. But if we want to prove that a specific number, such as $\sqrt{2}$, has a decimal expansion, then it is almost always easy to find an integer n that is greater than that number, in which case we can do without the Archimedean axiom. So the main use of the Archimedean axiom is in getting us from the axioms for a complete ordered field back to a more concrete picture of them.

Second, let me turn to the definition of continuity. Here, surely, is one of the truly difficult concepts that a beginning student of analysis must grasp. To teach it, people often start with a hand-waving explanation of what a continuous function is—it is a function "whose graph you can draw without taking your pen off the paper"—and they follow it up with a bizarre definition that appears to have nothing to do with this intuitive idea. As if to emphasize that the intuitive idea and the formal definition are different, students are given examples of pathological functions, such as the function that is continuous at all irrational numbers and discontinuous at all rational numbers, and are encouraged to be suspicious of their intuition and use the rigorous definition instead.

Does it have to be this way? I would contend that it does not because there is a much better intuitive description of what continuity is, one that leads directly to the rigorous definition. It concerns limited-accuracy measurement.

Suppose that a car is being driven along a flat road with its engine switched off and its brakes off as well, and we want to predict where it will be when it comes to rest. To help us, we are given full details of the frictional forces that it is subject to, and, crucially, we are told how fast it is going. Obviously, we are not given the speed as a real number, since we cannot know it exactly; rather, we are given an *approximation* to its speed, accurate to a few decimal places.

Because we are not given the exact speed, our prediction cannot be expected to be exactly accurate either. Is this a problem? In practice, no, because knowing the final position to a good approximation is good enough for practical purposes.

But can our prediction even be expected to be *approximately* correct? Most people feel instinctively that it can. Indeed, they somehow sense that the more accurate the initial data, the more accurate the prediction. Turning things around, they find it intuitively clear that if one insists on a certain level of accuracy for the prediction, then this accuracy can be achieved, provided the initial data are known sufficiently accurately.

Now let us vary the experiment slightly. This time the car is approaching a small sloped bridge. There are therefore three possible outcomes: It can come to rest beyond the bridge, it can come to rest at the top of the bridge, or it can go part of the way up the bridge before rolling back and coming to rest on the same side of the bridge that it is on at the moment. What it cannot do is come to rest on the parts of the bridge where there is any noticeable slope.

Suppose that the maximum speed that does not cause the car to go over the bridge and down the other side is 10 miles per hour. And suppose that the car is going at precisely 10 miles per hour. Then no matter how accurately we measure the speed of the car, we cannot be sure whether it will come to rest on the top of the bridge or over on the other side. Why is that? Because our measurement tells us that the speed of the car in miles per hour lies between $10 - a$ and $10 + b$ for two positive numbers a and b, and within that interval there are speeds where the car goes over the bridge and speeds where the car comes to rest on top of the bridge. Thus, however accurate our measurement is, we cannot even say *approximately* what the final position of the car will be.

What is the mathematical difference between the two variants of the experiment? In the first case, the final position of the car depends continuously on its initial speed, and in the other case, the dependence is discontinuous. This difference is easy to see intuitively, and if one tries to explain in detail the thoughts behind one's intuition, then one is led naturally to the conventional definition of continuity. This ease is not true of the graph-drawing intuition.

Here, briefly, is another way that one might explain to a high school mathematician what continuity is. Just ask them the value of π^2. They quickly ask you whether they are allowed to use a calculator, to which you reply yes. So they key in π and then press the x^2 button. The answer comes up: 9.8696044. You then express surprise: Is it really true that π^2 is a rational number? No, they explain, but π^2 is an infinite decimal, so the best they can do is give you the first few decimal places. Now you ask how they know that they have worked out π^2 to the first few decimal places. After all, the number they squared was not π itself but an approximation to π, such as 3.1415926. They probably protest that if the approximation to π is good enough, then the approximation to π^2 is good too. And they will have formulated for themselves a statement that is close to asserting the continuity of the function x^2. You can follow this up by asking them how accurately you would need to know π if you wanted to know π^2 to 100 decimal places. In that way, they would, without realizing it, be proving the continuity that they had just asserted.

Note that at no point in this conversation would they need to mention an epsilon or a delta, and yet their conception of continuity would not be importantly different or less rigorous than the standard one taught in universities.

My third example was the intermediate value theorem. This is another result that puzzles people because it seems obvious. However, if we put together the discussion about why 2 has a square root with the discussion of why one can feel confident that keying π into a calculator followed by x^2 gives one a good approximation to π^2, then we have everything we need for a rigorous proof of the intermediate value theorem in that special case. Furthermore, the resulting proof is close to the proof of the intermediate value theorem by repeated bisection. (It is not quite identical, because the theorem is slightly simpler if the function is monotonic.)

I firmly believe that it would be helpful if more could be done to show that the conventional treatment of basic real analysis is related to, and flows from, the kinds of intuitions that a high school mathematician already has about real numbers and functions defined on the real numbers. Of course, there is already a lot to teach, so fitting more into the curriculum may be difficult. But there are no such practical considerations for writers of textbooks. Unfortunately, there are still many textbooks, including newly published ones, that make no attempt to bridge the gap between high school and university mathematics. This could, and in my view should, be changed.

Mathematics Teachers' Subtle, Complex Disciplinary Knowledge

BRENT DAVIS

What mathematical competencies must a teacher have to teach the subject well? This question has proven difficult to investigate (1). A current view is that teachers' knowledge of mathematics "remains inert in the classroom unless accompanied by a rich repertoire of mathematical knowledge and skills relating directly to the curriculum, instruction, and student learning" (2). Unfortunately, there is no consensus on which "knowledge and skills" might activate teachers' inert knowledge. Two perspectives prevail, neither with a research base that enables strong claims about practice. The majority of current studies focus on explicit knowledge of curriculum content and instructional strategies. Such knowledge might be assessed directly through observation, interview, or written test (2), with a parallel research emphasis on the formal contents of teacher education programs [e.g., (3)]. A second school of thought, presented here, is that the most important competencies tend to be tacit, like skills involved in playing concert piano, learned but not necessarily available to consciousness.

Tacit Knowledge

Teachers' tacit knowledge includes many instantiations invoked to introduce and elaborate concepts, e.g., analogies, metaphors, and applications. Such instantiations are important in early mathematics learning. Teachers in high-performing jurisdictions, such as Hong Kong and Japan, were roughly twice as likely as U.S. teachers to invoke varied interpretations of concepts (4). Analogies can be useful, provided that novices have access to sustained interpretive assistance (5, 6). Despite considerable research on instantiations of grade-school mathematics concepts (7, 8), the topic has not been systematically incorporated into teacher preparation. This gap raises interesting issues, which can be highlighted through popular understandings of multiplication.

If you were to ask English-speaking students, parents, or teachers to define multiplication, you would likely hear a regular pattern of response. Most will insist that multiplication is repeated addition and/or a grouping process. This definition works for natural numbers, but it begins to break down as early as the middle grades. How, for example, does one add ⅝ to itself ¾ times, d to itself π times, or -2 to itself -3 times?

Notably, the importance of alternative interpretations has not been lost on the authors of classroom resources. Grade-school textbooks used in the English-speaking world typically invoke about a dozen distinct instantiations of multiplication by eighth grade: e.g., stretching and compressing a number line, array making, area generation, and a linear function. However, it is not clear that such text-based exposure translates into deliberate in-class examination of diverse interpretations and their entailments. Lacking such attention, learners may miss opportunities to develop robust and flexible understandings.

Learning Mathematics

The issue is not with the concept of multiplication itself. Within formal mathematics, multiplication is logically consistent and well defined, albeit the definition continues to evolve with the emergence of new number systems and other conceptual developments (9, 10). Nor is the concern with multiplication in particular. The same argument for diverse instantiations can be made for addition, number, equality, and so on. For educators, the issue is more the dynamics of learning mathematics than the structures of formal mathematics. Young learners' understandings are anchored to narrow, idiosyncratic bands of experience and interpretation. How might they be helped to make sense of unfamiliar situations in ways that are mathematically sufficient but neither overly rigid nor overwhelmingly complex?

This question can be difficult to answer for the expert, who moves effortlessly among instantiations, making choices fitted to the subtleties of the situation (11). To illustrate, Figure 1 shows four different interpretations of multiplication. Although there are some overlaps, moving

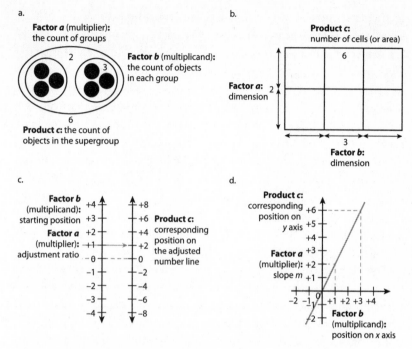

Figure 1. 2 × 3 as viewed through four interpretations of multiplication: (a) repeated grouping, (b) grid or area making, (c) number-line stretching or compressing, and (d) a linear function ($y = mx$).

among these four involves some conceptual leaps. It is not simply that the images are different; the actions that are mapped onto the concept of multiplying (clustering versus array making versus compressing versus sloping) are experientially distinct. Different mappings open up and shut down different interpretive possibilities (*12*). Collected together, they offer complex possibilities not present in any single instantiation (*13*).

The expert's ease of selecting and blending interpretations helps mask the subtle complexity of the knowledge. This subtle complexity makes it difficult to research the nature, extent, and relevance of teachers' tacit disciplinary knowledge. Investigations into the relation between teachers' formal mathematical preparation and their effectiveness have regularly shown little or no correlation between courses in formal mathematics taken by teachers and the performance of their students on standardized tests (*14*). This lack of correlation has troubled mathematics educators for decades, but from the perspective of tacit knowledge, there seems little reason to expect a strong relation. Teachers' university courses in mathematics typically focus on completed ideas, wrung free of the messiness and complication involved in coming to a new insight. Teaching young learners, in contrast, is largely about drawing logical consistency from diverse instantiations. Humans are not principally logical creatures, but analogical, our capacity for logic reliant on the tendency to make connections (*15*).

This emphasis on associative learning highlights a distinction between an expert's knowledge and a teacher's knowledge. Whereas it is the mathematician's task to pack insights into tight formulations (theorems, formulas, and so on), it is the teacher's task to unpack (*16*). This insight led to the notion of "profound understanding of fundamental mathematics" to describe the necessary knowledge for effective teaching (*17*).

The descriptor "fundamental" may be antithetical to researching the complexity of teacher knowledge. It suggests primary principles—basic building blocks—that can be identified, cataloged, transmitted, and tested. As the example of multiplication shows, it is not clear that instantiations of concepts operate as fundaments. They appear more to work as agents in an ever-evolving system. It may be more productive to think in terms of "profound understanding of emergent mathematics," the knowledge needed by teachers is more than a well-cataloged set of basics. As with any domain of profound human competence, most of it is necessarily tacit (*18*). It is unlikely that an individual could be consciously aware of the ranges of interpretations that might be invoked for the broad array of concepts covered in school mathematics.

Thus, teachers' mathematics might be more productively viewed as a learnable disposition rather than an explicit body of knowledge. Teachers' attitudes toward excavation and creation of instantiations may be as important as their backgrounds in formal mathematics. Such a disposition is learnable to some extent—by, for example, involving teachers in identifying useful instantiations, investigating their utility, combining them into more powerful interpretations, and using new insights to inform practice (*19*). Although still in early stages, research indicates that focusing on usually tacit knowledge can have immediate, significant, and sustained effects on teachers' knowledge of mathematics, perspectives on learning, and classroom practices, as well as student engagement, understandings, and attitudes (*13*).

Research into tacit knowledge is also difficult in that this focus conflicts with deeply entrenched beliefs about mathematics and learning. For centuries, school curricula have been developed around the assumption that human learning is a principally logical process (*20*), contributing to programs of study that consist of parsed and sequenced concepts. Directed linear movement through series of concepts may be incompatible with the goal of deep understanding, given the complex, evolving, networked structures of personal understandings of number, equality, addition, multiplication, and so on (*11, 15*).

Assumptions about fundamentals and linear progress have supported an approach to curriculum and testing that contributes to (and perhaps relies on) narrow, rigid definitions. Among alternatives is a conception that invokes an evolving network (versus rigid hierarchy) image for a knowledge domain. This alternative shifts emphases to seeking out associations and crafting elaborations (e.g., "How can we think about multiplication here?"). A network approach might be facilitated by new media technologies that enable, e.g., hyperlinks, ongoing revision, and collective processing. Attending to inherent complexities of mathematics requires different curriculum structures and teaching practices than those that prevail in public schools.

New structures and practices might help improve attitudes toward the discipline. Currently, in the middle grades, there are proliferations of meaning for many already-defined concepts, e.g., addition, multiplication, number, and equality. Perhaps these "explosions" of meanings, coupled to increasingly abstract applications, might contribute to the common cocktail party confessions, "I was good at math until grade 6" or, more troubling, "I liked math until grade 6." A learner trying to understand, faced with sudden complication and little interpretive assistance, might begin to dislike the subject matter. The alternatives to deep comprehension—rote memorization and routinized application—make for neither an engaging mathematics (*21*) nor one well suited to emerging needs.

On that point, approaches to how teacher knowledge is studied appear to be coevolving with perspectives on why mathematics is taught in the first place. Schools have long emphasized the development of technical competence, which was an obvious need in an industrial economy. In a knowledge-based economy, the development of conceptual fluency is of increased importance and has been the focus of major initiatives in school mathematics (*22*). Emerging research into the subtlety and complexity of teachers' knowledge not only reveals that these initiatives have fallen far short of their lofty goals, it also may offer an important route to achieving them.

References and Notes

1. D. L. Ball, H. C. Hill, and H. Bass, *Am. Educ.* **29**(1), 14 (2005).
2. J. Baumert *et al.*, *Am. Educ. Res.* **47**, 133 (2010).
3. W. H. Schmidt *et al.*, *Science* **332**, 1266 (2011).
4. J. Hiebert *et al.*, *Teaching Mathematics in Seven Countries: The Results of the TIMSS 1999 Video Study* (NCES 2003-013, National Center for Education Statistics, U.S. Department of Education, Washington, DC, 2003).
5. L. E. Richland *et al.*, *J. Exp. Child Psychol.* **94**, 249 (2006).
6. K. B. Zook, *Educ. Psychol. Rev.* **3**, 41 (1991).
7. L. D. English, Ed., *Mathematical Reasoning: Analogies, Metaphors, and Images* (Erlbaum, Hillsdale, NJ, 1997).
8. G. Lakoff and R. E. Nuñez, *Where Mathematics Comes From: How the Embodied Mind Brings Mathematics into Being* (Basic, New York, 2000).
9. B. Mazur, *Imagining Numbers (Particularly the Square Root of Minus Fifteen)* (Farrar, Straus and Giroux, New York, 2003).
10. F. J. Swetz, *From Five Fingers to Infinity. A Journey through the History of Mathematics* (Open Court Publishing, Chicago, 1994).
11. A. Sfard, *Thinking as Communicating: Human Development, the Growth of Discourses, and Mathematizing* (Cambridge Univ. Press, New York, 2008).
12. J. A. Kaminski *et al.*, *Science* **320**, 454 (2008).
13. B. Davis, in *Proceedings of the 34th Conference of the International Group for the Psychology of Mathematics Education (PME)*, M. M. Pinto, T. F. Kawasaki, Eds., Belo Horizonte, Brazil, 18 to 23 July 2010 (Belo Horizonte, Brazil, 2010), vol. 1, pp. 63–82.
14. Recent research suggests a more complicated relation than has been assumed in earlier studies. See (*2, 3*).
15. G. Lakoff and M. Johnson, *Philosophy in the Flesh: The Embodied Mind and Its Challenge to Western Thought* (Basic, New York, 1966).
16. D. L. Ball and H. Bass, in *Proceedings of the 2002 Annual Meeting of the Canadian Mathematics Education Study Group/Groupe Canadien d'Étude en Didactique des Mathématiques (CMESG/GCEDM)*, E. Simmt and B. Davis, Eds., 23 to 27 May 2008, Sherbrooke, Canada (CMESG/GCEDM, Edmonton, Canada, 2002), pp. 3–14.
17. L. Ma, *Knowing and Teaching Elementary Mathematics: Teachers' Understanding of Fundamental Mathematics in China and the United States* (Erlbaum, Hillsdale, NJ, 1999).
18. M. Polanyi, *The Tacit Dimension* (Doubleday, New York, 1999).
19. B. Davis, *Math. Teach. Middle Sch.* **14**(2), 86 (2008).
20. W. Schubert, *Curriculum: Perspective Paradigm and Possibility* (Prentice Hall, New York, 1985).
21. L. D. English, Ed., *International Handbook of Research in Mathematics Education* (Routledge, New York, 2008), chaps. 5 and 6.
22. National Council of Teachers of Mathematics (NCTM), *Principles and Standards for School Mathematics* (NCTM, Reston, VA, 2000).

How to Be a Good Teacher Is
an Undecidable Problem

ERICA FLAPAN

I began teaching my own classes when I was in graduate school. At that time, I never gave much thought to the question of how to be a good teacher. I lectured, following the book, interacting with the students, explaining the material step-by-step, and working out sample problems. The students seemed to appreciate my energy, enthusiasm, clarity, and willingness to answer their questions, and that's all there was to it. I continued teaching quite happily in this manner throughout graduate school and two postdoctoral appointments.

Then I got a tenure-track job at a liberal arts college and suddenly began getting mediocre teaching evaluations. It wasn't that my evaluations were that bad. They just weren't as good as they had been when I was a graduate student and a postdoc. It seemed that the students at a liberal arts college had greater expectations for their professors than the students at a university had. The one comment that kept appearing in multiple evaluations was that I followed the textbook too closely. I was baffled by this complaint, since it had never occurred to me to do anything other than follow the textbook. I did different examples than those in the book, but of course the structure, content, and organization of the material came from the book. Where else could it come from? I was impressed that there were math professors who were able to give lectures without following a book. I was sure I would never be able to do that.

In the meantime, in the 1980s, the Calculus Reform movement developed as a response to the realization that students across the country were doing poorly in calculus. Although few of my students failed calculus, I was determined to take advantage of the flurry of articles being published on new approaches to teaching to learn to be a better teacher.

I read about different teaching methods, each of which worked wonderfully for the author of the article. I also talked to popular teachers that I knew of at various colleges. Whoever I talked to was just as eager to convert me to their pedagogical approach as the articles were. However, the different techniques I was hearing about were inconsistent with one another, and all were opposed to the traditional lecture-from-the-book style that I was using. I decided I should just pick an approach from those I had read about and try it myself.

Since my students complained that I was following the book too closely, I decided that the first thing I would try was to get the students to read the book themselves. After endless searching, I found a calculus book that was well written and seemed student friendly. I told the students that my lectures would not repeat what was in the book, and I began each class period by asking the students questions about what they had read. When nobody would respond to my questions, I decided to make the students hand in their responses in writing. However, as I read their responses, I realized that the students really weren't understanding what they were reading. I decided that I had to go over the parts of the text that they had completely misunderstood. Gradually, I began going over more and more of the material in class. By the end of the semester, I was convinced that the students got a better understanding of the material from my lectures than from their quite serious attempts to read the book. If my goal was for them to understand the material, shouldn't I be doing all in my power to help them do so? I could have tried variations on this method, but instead I sighed and decided to try something else.

My next attempt at innovation was to use an inquiry-based learning method to get the students to figure out the concepts of calculus themselves. I wrote detailed worksheets that would lead the students to understand the concept of the derivative, as well as to conjecture the product rule, the chain rule, and so on. During class, the students worked on my worksheets in groups and I walked around helping them as they struggled with the problems. This method motivated them to think about the concepts, and as a result they seemed to end up with a better understanding of what calculus is really about. Yet using this method took me about twice as long to cover each topic as when I taught the course by lecturing. Not only was I unable to cover all of the material that was required for Calculus II, but also, with so much class time spent on developing the concepts, the students were not getting much exposure to complex applied problems. When the students had completed my Calculus I class, I felt that they were neither ready for Calculus II, nor were they prepared to do the types of applied problems that they would see in their physics or economics classes. Yet, they had been well prepared in both of these areas with my traditional lecturing method. With a sigh, I again concluded that this method had not worked as well for me as it had for the people who had written the articles that I had read.

I considered, and then quickly rejected, various other methods of teaching calculus. For example, I could have the students use Mathematica or MatLab to experiment with different concepts in calculus, but this method seemed no better than my worksheets (which used calculators), and the thought of dealing with various computer problems that the students might have did not appeal to me. I considered a variety of calculus reform textbooks, but none seemed that much better than the textbook that we were using, and some seemed much worse.

The most popular teachers that I knew of would intersperse jokes or stories in their lectures to keep the students awake and interested. I had a colleague who put on mathematical skits in class and sang original songs about mathematics that the students loved. I ended up collaborating with him on a number theory textbook that interspersed funny stories and jokes with the material. Still I couldn't sing, act, or even remember a joke while I was teaching. The only times that the students laughed during my classes were when I did something funny by accident. Like when I would be erasing the top of the board and the eraser would fall on my head, leaving a rectangular white chalk mark in my black hair. Or the time when I grabbed a pole (meant to open the windows) to point at the ceiling as I explained how opposite walls in the room could be glued up to make a 3-torus. After hearing muffled laughter, I realized that the stick had caught on the bottom of my skirt and pulled it up higher and higher each time I pointed. Let's not even mention the trouble I got myself into trying to talk about the importance of balls in metric spaces. So it wasn't that my students weren't laughing in class. I just couldn't imagine how to be entertaining—on purpose.

Becoming frustrated with my attempts at innovation in elementary classes, I decided to try reforming my upper division classes. I knew that some students were only interested in mathematics because of its applications. I had heard that this phenomenon was particularly true for women students, but I didn't believe it since I had always loved pure math. In any case, I decided that I should try to integrate some applications into one of my pure math courses. I picked my Introductory Analysis class because this was the class that students had the most trouble with. After extensive searching, I adopted an analysis book that contained applications.

One of my goals in teaching Introductory Analysis was to teach students to prove all of their assertions rigorously without asserting that anything was obvious. Yet, once I began lecturing on the applied material, I realized that those sections of the book avoided all of the epsilons and deltas, so as not to get bogged down. I could supply the details of the proofs myself, but that would make the material seem tedious and boring. On the other hand, I didn't feel comfortable saying to the class, "It's okay for me to skip all the rigor when I'm doing applications, but it's not okay for you to be equally nonrigorous when you are doing your homework." In addition, as a pure mathematician myself, I still wanted to try to convince the students that results like the Bolzanno– Weierstrass theorem are beautiful independent of their applications. I sighed and decided that this idea really wasn't working that well either.

Another idea that I had read about was to motivate mathematics majors by bringing research into the classroom. As a topologist, I thought the best place to try this would be in my topology class. My research is on knot theory, 3-manifolds, and graphs embedded in 3-space. These topics did not fit neatly into the topics on my syllabus, and I didn't want to wait until the last few weeks of the semester to tell the students about research. So I decided to spend the first 10 minutes of every period teaching the students some knot theory. Several students thought this was really cool and would eagerly come to class excited to learn about it and tell me their ideas. However, most students realized that this material was not central to the course and preferred to sleep in and thus began arriving late to my 9:00 a.m. class. I tried moving the knot theory to the end of the period, but found that this gave some students an excuse to doze off after I had finished covering the "important" concepts of the class. Of course, "important" was defined as what would be on the exam. Around this time, I got e-mail messages from several alums who were having a hard time in math graduate school. After a big sigh, I gave up on knot theory so that I would have time to cover topics that I thought would better prepare my current students for a graduate course in topology.

About this time, I had a baby. While I was pregnant, I had read books about child rearing and talked to more experienced mothers about how to be a good mother. However, once I began parenting, I realized that the ideas I had read about actually didn't fit with my personality or the relationship I wanted to have with my daughter. I put the books back on the shelf to use as a reference if I became desperate (which I never did). Instead, I decided to trust my own instincts as a parent, just as I did in my relationships with my friends and my husband. This method seemed more genuine than trying to follow a philosophy of parenting that didn't come naturally to me. As I became more confident in my parenting, I tried to think about what made me a good parent. I decided that in addition to loving my daughter, the key to being a good parent was always listening to her with empathy and respect, and just being myself.

This idea led me to the realization that the same was true for teaching. Being a good teacher was not about using a particular technique or philosophy of teaching. Each one of the techniques that I had tried had worked wonderfully for some people, and I could probably make it work for me if I put in a little more effort. But, a method that didn't fit with my personality, mathematical interests, or style would never feel completely natural to me. To be a good teacher, all I had to do was listen to the students with empathy and respect, and then just be myself. In fact, I realized that the ability to listen to students and see the material from their point of view was my greatest strength as a teacher. When students asked questions, I was able to pinpoint what they didn't understand. And when students told me about what they liked and didn't like about math, I was able to help them set career goals and figure out what they needed to do to achieve them.

After I decided that there was no algorithm for how to be a good teacher, I considered whether I should go back to any of my previous attempted innovations to see if I could make them work for me. But instead I decided that outside of the classroom it would be better to spend my time mentoring current and former students at all levels, since this seemed like the most important gift that I had to give. In the classroom, I went back to giving clear interactive lectures as I had before, using no specific technique or philosophy that made my lectures better than those of any other professor who gave clear lectures.

By this time, I had been teaching for long enough that I had used various textbooks. My lectures now incorporated what I learned from each of the books I had used. So I no longer got complaints from students that I was following the book too closely. This situation wasn't the result of a decision to try not to follow a textbook. It just happened naturally with age and experience. In addition, I gradually developed a few tricks of my own. Here are some examples:

Trick 1: Instead of spicing up my lectures with a joke or a song, in the middle of each period I invite a student to ask an "irrelevant question." These questions can be anything the student wants to ask, and hence range from questions about careers in math to questions about my personal history. They can even include some creative questions like: "What advice would you give someone who wants to date a mathematician?" and "Would you rather marry Riemann, Cauchy, or Euler?" I was asked both of these questions last semester.

Trick 2: To motivate students to do hard homework problems, I make these problems worth between $\frac{1}{2}$ and 2 extra credit points that I will add to their next exam score. Often this means a student will only get $\frac{1}{10}$ of a point for an incorrect attempt on such a problem. So this does not substantially affect student grades, though students feel like it does. The students are eager to try these problems because they feel like they have nothing to lose, and possibly something to gain. I use this trick in my precalculus-level Problem Solving course to get the students to work on lengthy applied homework problems that take up to a page to state. The students never complain about how hard these problems are. At a more advanced level, I use this trick in my analysis class to get students to focus on writing proofs in a rigorous style. I give the students extensive criticisms of their extra credit proofs without them getting upset because they are still racking up fractions of extra credit points.

Trick 3: For upper-level classes with challenging weekly homework sets, some students ask me for hints without putting much effort into the problems. To deal with this problem, I have created two hints for every problem. The first hint is somewhat vague, and the second hint has more detail. To get the first hint, a student needs to show me that he or she has worked on the problem. A student cannot receive the second hint unless a day has passed since receiving the first hint and the student shows me that he or she has thought about the first hint. In particular, students who start working on the homework the day before it's due never get more than one hint. Although I don't always remember things students have told me several days or weeks before, I have no trouble remembering whether a particular student has already appeared once that day to ask for a hint. So there is no need to keep track of how long ago I gave out a hint.

TRICK 4: To force upper-level students to pay attention and participate actively in class without letting the strong students dominate the discussion, I do proofs "in the round." That is, I go around the room asking each subsequent student to supply the next sentence or small step in a proof. Something as simple as stating the relevant definition or rewording what we are trying to show is fine. If a student can think of nothing to say, I ask leading questions until the student says something that I can write on the board. I don't let students answer out of turn. At first the students find this method stressful. But by the end of the semester, the students get used to it and feel that it helps them stay focused and really understand the steps of the proofs.

This gives you an idea of the sorts of tricks I have developed. I find them useful, but they are still just tricks. They don't make me a better teacher. And they may or may not work for other teachers, with other types of students, or at other types of institutions.

For some people, the satisfaction of teaching comes from their enthusiasm about the material. For some, it comes from the opportunity to put on a performance. For me, the satisfaction of teaching comes from getting to know and mentor a diverse group of wonderful students. I never get tired of teaching the same class over and over because the students are always different and I always find them interesting. I love to stay in touch with my students well beyond my classes, not only to continue to mentor them as they develop over the years, but because many become friends with inspirational careers and ideas that I can learn from.

I won't tell anyone that they should teach like me, just as I won't tell anyone that they should parent like me. You have to care about and respect your students (and your kids), be clear in your lectures, and stay excited about the material. Beyond this, you should use your own experience and self-awareness to tailor your teaching style to your strengths and interests. As long as you do these things, there is no right or wrong way to teach.

In computational complexity theory, an "undecidable problem" is a problem for which it is impossible to construct a single algorithm that leads to a correct answer in all cases. It is pretty clear to me that there is no single pedagogical method that always leads to maximal student learning. Perhaps we should exert less effort trying to decide the undecidable and instead spend more time enjoying our students, our classes, and mathematics itself.

Acknowledgment

The author was supported in part by NSF Grant DMS-0905087.

How Your Philosophy of Mathematics Impacts Your Teaching

BONNIE GOLD

"My philosophy of mathematics? I don't have one! I'm a mathematician, not a philosopher. I leave philosophical questions to the philosophers." Maybe. Or perhaps you are among those mathematicians who *are* interested in the philosophy of mathematics. Whatever your attitude toward the philosophy of mathematics, when you teach mathematics, you do in fact take, and teach your students, positions on philosophical issues concerning mathematics. If you do not think about them, then you probably acquired your positions from *your* teachers when they imposed them on you without discussion. Furthermore, you may find, if you do examine the positions you are taking, that they contradict each other or disagree with positions you would say are obviously true.

Many mathematicians' distaste for philosophy of mathematics comes from one of two sources. One is our preference for questions that, if worked on seriously, eventually receive definitive answers, as opposed to questions (as in philosophy) that, at best, clarify what the issues and alternatives are. The other is the detour the philosophy of mathematics took into foundational issues starting in the second half of the nineteenth century. This detour lasted until about 1975, when work on questions beyond foundations resumed. The detour led to a substantial growth in the area of mathematical logic, and some consensus on what was *not* going to be solved. (For example, we are *not* going to be able to show in a finitistic manner that mathematics is consistent.) However, most mainstream mathematicians lost interest after Gödel's work.

I use a rather broad definition of philosophy in this article: It includes anything involving our attitudes toward mathematics. I am not trying here to convince you to start working in the philosophy of mathematics.

But I do hope to make you aware of a range of issues that we end up taking a stand on when we teach a relatively standard calculus or introduction to proof course. Many of these issues are closely related to issues that beginning students are confused by. Perhaps a lack of clarity on philosophical matters adds to students' confusion.

The Nature of Mathematical Objects (Ontology)

Ontology is one of the hottest issues among philosophers of mathematics. But most mathematicians simply do not care whether mathematical objects are abstract or concrete, fictions, communally created, or real. The old saw that the mathematician is a Platonist when he is working on mathematics and a formalist when discussing philosophy is often true. As Errett Bishop replied, when a philosopher questioned him about the distinction between numbers and numerals, "I identify a number with its numeral." (The philosopher was foolish enough to follow up with, "So how do you distinguish the numeral 2 from the numeral 3? They are both curly," to which Bishop replied, "The numeral 2 is divisible by 2; the numeral 3 isn't." This interchange occurred at Cornell; I was in the audience.) We do not care *what* the number 2 is; what we care about is that it has certain properties. So most mathematicians do not care if you identify 2 with $\{\emptyset, \{\emptyset\}\}$ or with $\{\{\emptyset\}\}$ or don't identify it with a set at all. What is important to us is that there be no ambiguity about how mathematical objects behave. Whether you identify 2 with one of these sets, or do not make any such identification, $2 + 3$ still equals 5, and 2 is still a prime number. This is what matters to us.

NUMBERS

Our students, however, have difficulties with some mathematical objects. Virtually all of them agree that $1/3 = 0.3333\ldots$, but many are uncomfortable with $1 = 0.9999\ldots$. They want $0.9999\ldots$ to be the number "just to the left of 1." And we *have* made a choice when we tell them that $1 = 0.9999\ldots$: We have decided that there are no infinitesimals in the real numbers. (On the other hand, if there *are* infinitesimals in the real numbers, then $0.9999\ldots$ is ambiguous: How far into the infinite does "$\ldots$" go?) For that matter, students often believe that $1/3 = 0.3333333333$ (without the "$\ldots$") because that is what they see on their calculators when they punch "$1 \div 3$." To add to the confusion, when they then multiply this result by 3, their calculator gives them 1, although if they punch 0.3333333333 in directly and multiply by 3, they get 0.9999999999. These confusions present an excellent opportunity to discuss a range of issues, such as how the real numbers were developed, how mathematics develops in general, a range of paradoxes, and several questions concerning approximation.

Sometimes the difficulty, is caused by our taking a word that has a meaning in everyday speech and changing its meaning slightly for use in mathematics. For example, calculus students often have difficulty with the idea that a function can achieve its limit. This difficulty may be a combination of the meaning, in English, of "limit" (though many of them, when they drive, not only achieve, but exceed, the speed limit) together with the fact that, in our definition of $\lim_{x \to c} f(x)$, we don't let $x = c$.

Many of Zeno's paradoxes relate to limits or to what are called completed infinities. It can enhance students' introduction to infinite series to consider the question of whether Achilles could ever beat the tortoise in his race. After all, the tortoise had a head start. By the time Achilles got to where the tortoise was when Achilles started, the tortoise had moved a bit further on. Another of Zeno's paradoxes related to calculus: If an arrow is always at some place at every time, then how can it be moving? These questions all relate to issues of limits and the nature of real numbers.

EQUALITY

There are often subtle questions, both philosophical and mathematical, about when two mathematical objects are the same. For example, is $10^{\log(a)} = a$? Yes, when a is positive, but the left side is meaningless when a is negative or zero. What kinds of mathematical objects are under consideration at a given time has a lot to do with whether or not they are equal. Also, two mathematical objects that arose in different contexts often turn out to have the same properties and are eventually identified as being the same object. There is an interesting discussion of this issue in Mazur [7].

Functions play a very important role in mathematics. As students begin their university studies, their notion of function is usually a formula: $x^2 + 5x, 2x/(x - 2), \sin x$—what our computer algebra systems insist is an expression. Most calculus courses (starting with the calculus reform movement) work on broadening the concept. Many of us choose to teach undergraduates that a function is like an input–output machine that, given any element of the input set, produces a unique element of the output set. An exercise I use, borrowed from Ed Dubinsky [8], is to give students in my introduction to mathematical reasoning class a range of situations (such as a statement like "Monmouth has a very good basketball team" or a graph that does not pass the so-called vertical line test). I then ask them, wherever possible, to come up with one or more functions relevant to that situation. This is quite a different approach from giving students the standard formal definition of a function (a subset of the Cartesian product of two sets A and B, such that $(\forall x \in A)(\exists y \in B)((x, y) \in f)$ and $(\forall x, y, y')(((x, y) \in f \wedge (x, y') \in f) \Rightarrow y = y'))$.

The first way of working with students fits well with a Platonist or social constructivist view of mathematics. Platonists view mathematical objects as real objects that we can explore as if we were exploring a new country. Social constructivists view mathematical objects as constructed by the mathematical community. For them, engaging students in an exploration of functions is part of bringing them into this community. The formal definition approach fits better with formalism than with either of these more popular views. Certainly *eventually* we want all budding mathematicians to be comfortable with the formal definition for efficiency of communication. But only people who are formalists about pedagogy (and not even all of them) believe it to be the best way to introduce the concept.

Formalists and nominalists often find themselves speaking in a way that appears inconsistent with their beliefs. You may believe that, since the only things that exist are physical, there are not really any mathematical objects. But watch out! In class, when discussing continuity versus differentiability, you may find yourself saying, "Now let's construct a continuous, nowhere differentiable function."

In general, our understanding of a mathematical object grows as we study more mathematics. Our concept of function, as a map from real numbers to real numbers, is extended to functionals such as the derivative, to homomorphisms of assorted types, and so on, sometimes even to objects, in category theory, that no longer meet the formal definition of function because they are no longer defined on sets.

Mathematical Truth

As a student, one of the most attractive features of mathematics for me was that a mathematical problem had just one right answer. Furthermore, that answer was as open to inspection and verification by a beginning student as by a teacher. However, as I grew a bit more sophisticated, I learned that there are mathematical questions with several correct answers (not inconsistent ones, but of a range of depth), depending on the context in which the question is set. Also, often when it seems that a problem has been answered, it is later possible to relate it to other mathematical areas and, in doing so, raise variations on the question that have not yet been answered. (There is a nice article by Phillip Davis [4] on this.)

THEORIES OF TRUTH

When we teach certain standard theorems of calculus (such as the extreme value theorem or the intermediate value theorem), we take a stand against intuitionism/constructivism. In particular, these theorems have constructive versions (slightly more complicated to state) that, classically, are equivalent to the theorems we teach. However, the standard versions found in most calculus or real analysis books are not theorems constructively. For example, a constructively true form of the intermediate value theorem is that, if $f : [0, 1] \to \mathbf{R}$ is continuous, with $f(0) < 0$ and $f(1) > 0$, then for each $\epsilon > 0$ there is an x in $[0, 1]$ such that $|f(x)| < \epsilon$. Or, if you add to the standard hypotheses that for every a and b in $[0, 1]$ there is an x in $[a, b]$ with $f(x) \neq 0$, the standard conclusion becomes constructively true. See [3, p. 59], for more variations.

Simply to say a certain mathematical statement is *true* involves taking a philosophical position. If you are a formalist, you say, rather than that "the theorem is true," that "it is a theorem within a given axiom system." For a substantial collection of philosophers of mathematics (called nominalists, or a subcategory, fictionalists: see Balaguer [1]), there are *no* mathematical truths, because there are no mathematical objects for them to be true about. The most you can say is that an assertion is "true in the story of mathematics," a story that the community of mathematicians builds, just as Conan Doyle built the story of Sherlock Holmes.

I've always been some kind of "realist" about mathematical objects and mathematical truth (though partly a social constructivist about our knowledge of them). Mathematical statements are true because they accurately describe mathematical objects. Yet as a young faculty member, when asked why $0! = 1$, I tended to give a formalist/ conventionalist answer: "Because we can make it anything we want, and we define it that way." Yet I really don't like this answer either pedagogically—"because I say so" is not an attitude I want to teach— or philosophically, since there are far better, if longer, answers that are consistent with my philosophy.

The Persistence of Truth

Mathematics appears to be almost unique in that, once a problem is solved, the original solution remains true forever (except, of course, in the relatively rare cases where an error is published). This is a phenomenon that is a challenge for some (nominalist, fictionalist, and some social constructivist) philosophies of mathematics. It is certainly a property students should be made aware of. In fact, students often fail to appreciate that a mathematical property they learned in one mathematics course remains true in the next one. As my Introduction to Mathematical Reasoning students are constructing their first baby number theory proofs about even and odd integers, they often forget basic algebraic properties. So I get many proofs in which, for example, since a and b are even and they are trying to prove that ab is even, they let $a = 2x$, $b = 2y$, and then conclude that $ab = (2x)(2y) = 2xy$.

Verification of Truth

For most beginning students, however, the main problem with mathematical truth is in our rather peculiar method of determining it, completely different from anything in their previous experience. In many real-world situations, "the example proves the rule." That is, giving one example may make it clear that the statement is, in fact, generally true, and when this is not the case, some kind of statistical inference from data is often used to establish truth. Not in mathematics: A general statement is only agreed to be true if we can give a proof. Although we insist on this in our proof courses, in lower-level classes, we often imply exactly the opposite. Rather than giving a proof, we may give just one or two examples illustrating a theorem, perhaps along with a counterexample when one of the hypotheses fails. I am *not* arguing against doing this. I do believe, at least in beginning courses, that proofs *are* appropriate when the result is counterintuitive (such as the product rule for derivatives), but not when a picture accurately illustrates the general situation. What I advocate is being more explicit about the role of proof and the role of examples in determining truth in mathematics. And there are quite a few roles for examples. First, examples are used to show that a general statement is false. Second, a good (fairly generic) example often *does* lead to a proof. Examples are also extremely helpful for understanding what a theorem is saying. They often also lead to an understanding of what is happening in a mathematical situation, which helps us generate a conjecture. Far too often, once our students have become convinced that an example cannot replace a proof, they stop using examples altogether.

This issue of examples also comes up in proof by cases. For students, doing a proof by cases seems a lot like doing a proof by giving a few examples, and the distinction needs to be made clear.

Mathematics and the Real World (Epistemology)

PICTURES, GRAPHS, AND DOODLES

The role of drawings in mathematics is getting increasing attention recently among philosophers of mathematics. Several books and articles (e.g., [2], [6]) have been written on the topic. Most of us enjoy the proofs without words that sometimes appear in assorted Mathematical Association of America journals. But I think most of us feel that these are not quite proofs. To turn them into what we would consider proper proofs, one needs to add statements such as "Given any triangle, with . . ." and turn the picture information into algebraic information to make sure that it is indeed sufficiently general.

On the other hand, pictures are efficient ways to communicate the idea of what is happening mathematically—and, depending on your philosophical viewpoint, that may be exactly what mathematics is about. Some of our students are not visual learners, but for most of our students, accompanying algebra with a picture makes the situation clearer. In any case, whatever you say to your students about pictures and their use in mathematics implies a philosophical standpoint. It thus is worthwhile to give some thought to what you believe pictures have to do with mathematics. If you are a formalist, until we have a formal system for using pictures, they are simply a device to help people who are visual develop formal proofs. For a Platonist, pictures may play an important role in how the physical beings that we are can have contact with the mind-, space-, and time-independent realm of mathematical objects. As a fictionalist, pictures may be part of "our story of mathematics"—the illustrations?

The Unreasonable Applicability of Mathematics

Another issue is the question of why mathematics is so applicable, particularly to problems other than those it was developed to solve. Again, this question poses problems for various philosophies of mathematics (some Platonist views and formalism, for example). Why should abstract objects or formal deductions have anything to do with the physical world? Students also need to be made aware of the dangers of applying mathematics to the world without a careful model-building process and of the dangers of extrapolating mathematical conclusions beyond the limits of the model. This warning gets back to the basic philosophical issue of being able to distinguish what we know from what we do not know.

Language and Logic (Semantics)

As we move through school as budding mathematicians, we learn the language of mathematics, and by the time we start teaching, it is second nature. But there are a number of ways that mathematicians' use of language differs from our students' precollege experience.

Logical Connectives

Susanna Epp has written eloquently about differences between everyday speech and mathematical use of logical terms such as "if . . . then . . .," "or," and "not" when combined with a quantifier (see, e.g., [5]). In everyday usage, "or" can be inclusive ("Would you like sugar or cream with your coffee?") or exclusive ("Would you like soup or a salad?"). It is most often used exclusively, whereas in mathematics it is almost always used inclusively. In everyday speech, we frequently use the "if . . . then . . ." structure when we mean "if and only if." ("You may watch television if you finish your homework.") In mathematics, this distinction is essential: A statement is equivalent to its contrapositive, which is important in several types of proofs, but *not* equivalent to its converse or inverse. Also, in mathematics, we say that $P \Rightarrow Q$ is true as long as P is false or Q is true. This is very counterintuitive to most students. They usually believe that "if pigs fly, all even integers greater than 2 are composite" is *false*, because pigs do not fly. Negations are also used differently: "Mom, everyone is going to the after-prom." "No, Jenny, everyone is *not* going to the after-prom—I just talked with Cindy's mom, and Cindy isn't going." Mathematically, the "not" belongs *before* the "everyone": "No, Jenny, *not* everyone is going to the after-prom. . . ." But that is not how we speak. We are also quite insensitive, in everyday speech, to the order of quantifiers, yet in mathematics $(\forall x)(\exists y)$ is *very* different from $(\exists y)(\forall x)$.

On most of these issues, mathematicians are united, but pedagogically we need to be aware of these notational issues when we are trying to initiate our beginning students. That is, we need to be sure we are all using the same "semantics." There are other logical issues on which we are not completely united, where you must take a stand.

OTHER LOGICS

First, of course, is where you stand on intuitionism/constructivism versus standard logic. Intuitionists only accept the law of the excluded middle in restricted circumstances, and certainly not in the situation where you want to prove that something exists by showing that its non-existence leads to a contradiction. You may feel that the law of the excluded middle—that for every statement P, either P or its negation (symbolized here by $\sim P$) must be true—is obvious. For an intuitionist or constructivist, however, you do not know this until you know *which* one is true. Fermat's last theorem became true when Wiles proved it. Before then it was neither true nor false, simply unknown. For very simple statements P, students will usually agree that either P or $\sim P$ must be true, but once a statement gets complicated with quantifiers and negations, this "truth" is less clear to them. Certainly in everyday language, it can be the case that neither P nor $\sim P$ is true: for example, one or both may be meaningless (e.g., "circles are green").

Students are often uncomfortable with the use of the law of the excluded middle in proofs by contradiction. It is bad enough that we tell them they cannot assume, in their proofs, the statement they are trying to prove. Now we tell them that one method of proof is to assume the negation of the statement they are trying to prove! It may help somewhat to tell them that, although most mathematicians currently agree that this is a valid method of proof, in some situations there are mathematicians who reject it.

Generally, there are more logics that people use than simply the two-valued logic of most mathematics textbooks: tense logics, modal logic, and so on. You certainly are welcome to stick with two-valued logic, but you thereby make a philosophical choice.

DEFINITIONS

Definitions have a different status in mathematics than in everyday language. Usually one learns words by ostension—that is, by seeing an example or having one described (say, a table, or the disease called shingles), and the person who is teaching the word attaches it to the example or situation. One usually uses definitions to distinguish between two words that have similar (but not identical) usages. A given object may satisfy a particular definition more or less well: Is a stool a chair? In most situations, a tree stump is not a chair, but it may function as one. In mathematics, definitions are used to carve out, with no ambiguity, certain collections of mathematical objects. Any statement about one of these objects generally can be replaced with the definition in proofs or examples. So, in some sense, giving a definition for a mathematical object says everything there is to be said about it. Of course, in other ways it certainly does not: One actually learns the meaning of a mathematical concept by working with examples as well as theorems involving it.

From a formal standpoint, however, the definition is all there is to the concept. Thus, if you teach a course in the traditional definition–theorem–proof mode, you essentially endorse formalism. (Of course, one can be a formalist about how one learns mathematics without being a formalist about what mathematics is—more on this distinction later.) In any case, some discussion with students about the role of definitions in mathematics (at least in courses for mathematics majors, and courses for teachers at any level) is helpful for them.

NOTATION AND ITS ABUSE

A related issue is symbolism. Much mathematics (for example, certain solutions of differential equations) was discovered by saying, "Let's assume that this notation works in this new situation and see if it leads us to a correct solution." (Some theoretical physicists similarly seem to use mathematical symbolism as magic.) Furthermore, much of students' high school mathematics consists of manipulating a long chain of symbols until it turns into what is required. Perhaps from this method, our students often use symbols as if certain properties automatically come with them, particularly distributivity and commutativity. Hence, they tend to move without thinking from $\sin(x + y)$ to $\sin(x) + \sin(y)$ (and similarly with most functions, whether linear or not).

This problem is not entirely our students' fault. Mathematicians are notorious for abuse of notation. We often use the same symbols in several different contexts (overloading the notation). For example, (a, b) can be a point in the plane, an interval on the real line, the greatest common divisor of a and b, a member of any Cartesian product, and so on. We use variables in a wide range of ways, expecting the student to figure out how it is being used from the context. Does a represent a range of values of an input variable for a function, the (finite number of) solutions to a given equation, an arbitrary real number that can be substituted in an identity, a parameter that we are keeping fixed for the time being while x acts as an input variable, a bound variable within some specific range, a free variable? Eventually, mathematicians get used to determining the meaning from the context, but in our freshman courses, some mention of context often clarifies a statement that otherwise mystifies much of the class. Again, a discussion of mathematical notation and mathematical symbolism requires some thought about these issues.

Teaching and Learning Mathematics
(Pedagogical Epistemology)

How one introduces new mathematical concepts—whether by giving a few examples first, or by asking questions or working on problems that bring out the need for the concept, or by giving the definition and a few theorems—is strongly related to your beliefs about how mathematical knowledge is acquired. And your belief in how mathematical knowledge is acquired is often related to what you believe mathematics is about: e.g., objects independent of us, socially constructed objects, or formal deductions from axioms. A formalist, for example, is likely to give axioms and definitions first.

How Do We Learn Mathematics?

The study of how we acquire mathematical knowledge is still in its infancy. We are learning that certain approaches work better than others, but there is certainly no coherent theory that is widely agreed on yet. We learned mathematics ourselves via a range of approaches and activities. Furthermore, what works well for those of us who go on to become mathematicians often does not work for the vast majority of students in our classes.

Usually the route by which a mathematical concept was originally developed is not how we teach it, once the particular area is well enough understood to be taught to undergraduates. Certainly, I have not heard anyone supporting the view that students usually should learn a concept in exactly the same way it was originally discovered. On the other hand, the mathematics curriculum broadly recapitulates the history of mathematics. There is a fairly popular view among educators that students should construct their understandings of mathematical concepts. However, if one teaches this way, one carefully guides students' explorations in the process of making these discoveries so that they do not have to take the centuries it took the human community to develop the mathematics.

Playing One Hand against the Other

Since many questions about how students learn mathematics remain unanswered, most of us—I certainly include myself—are rather conflicted in our beliefs about how mathematical knowledge is acquired. Without reflection, most people teach in the manner they were taught, rather than in a manner reflecting their belief of how knowledge is acquired. It is quite possible to be a formalist about mathematics—believe that mathematics is just a formal game played with symbols from a given set of axioms—and yet believe that to get students to learn to play this game effectively one should have them make certain kinds of constructions. Or one could be a social constructivist about how mathematical knowledge is developed by the community of mathematicians and still have students learn it formally as their entry ticket into that community.

But generally it does seem to make sense that, if you believe that mathematical knowledge is socially constructed, you would have your students socially constructing their own mathematical knowledge, at least to some extent. Similarly, if you believe that mathematics is, in a broad sense, about phenomena that are part of the world we live in (both physical and mental, say), then you are likely to have students approaching mathematics at least somewhat as they approach other sciences, by a certain amount of (guided) discovery. (You can hold both of these beliefs, by the way, as I do. I believe mathematics is in the world around us, waiting to be discovered, but that our mathematical knowledge, just like our knowledge of any other science, is constructed by the community of scholars.)

Conclusion

I hope I have convinced you that there is a substantial range of philosophical questions on which you take a position when you teach. By what we say in class we take a stand regarding what mathematical objects are, the role of definitions in mathematics, the kind of logical rules that are to be followed, and how mathematicians determine truth. Your students will benefit if you give some thought to these issues before you unconsciously take, and teach, a position.

Acknowledgment

An early shorter draft of this paper was presented in the contributed paper session, "The History and Philosophy of Mathematics, and Their Uses in the Classroom," at MathFest 2009. I would like to thank both the referees and the editor for their very helpful suggestions toward improving this article.

References

1. M. Balaguer, "Mathematical Platonism." B. Gold and R. Simons, eds., *Proof and Other Dilemmas: Mathematics and Philosophy*, Mathematical Association of America, Washington, DC, 2008, 179–204.

2. J. Barwise and J. Etchemendy, "Visual information and valid reasoning." W. Zimmerman and S. Cunningham, eds., *Visualization in Teaching and Learning Mathematics*, Mathematical Association of America, Washington, DC, 1991, 9–24.

3. E. Bishop, *Foundations of Constructive Analysis*, McGraw-Hill, New York, 1967.

4. P. Davis, "When is a problem solved?" B. Gold and R. Simons, eds., *Proof and Other Dilemmas: Mathematics and Philosophy*, Mathematical Association of America, Washington, DC, 2008, 81–94.

5. S. Epp, "The role of logic in teaching proof." *Amer. Math. Monthly* **110** (2003) 886–899. doi:10.2307/3647960

6. M. Giaquinto, *Visual Thinking in Mathematics: An Epistemological Study*, Oxford University Press, Oxford, U.K., 2007.

7. B. Mazur, "When is one thing equal to some other thing?" B. Gold and R. Simons, eds., *Proof and Other Dilemmas: Mathematics and Philosophy*, Mathematical Association of America, Washington, DC, 2008, 221–241.

8. D. Breidenbach, E. Dubinsky, J. Hawks, and D. Nichols, "Development of the process conception of function." *Educational Studies in Mathematics,* **23** (1992) 247–285.

Variables in Mathematics Education

SUSANNA S. EPP

Variables are of critical importance in mathematics. For instance, Felix Klein wrote in 1908 that "one may well declare that real mathematics begins with operations with letters" [3], and Alfred Tarski wrote in 1941 that "the invention of variables constitutes a turning point in the history of mathematics" [5]. In 1911, A. N. Whitehead expressly linked the concepts of variables and quantification to their expressions in informal English when he wrote, "The ideas of 'any' and 'some' are introduced to algebra by the use of letters. . . . it was not till within the last few years that it has been realized how fundamental any and some are to the very nature of mathematics [6]." There is a question, however, about how to describe the use of variables in mathematics instruction and even what word to use for them.

Logicians seem generally to agree that variables are best understood as placeholders. For example, Frege wrote in 1893, "The letter 'x' serves only to hold places open for a numeral that is to complete the expression. . . . This holding-open is to be understood as follows: all places at which 'ξ' stands must be filled always by the same sign, never by different ones" [2]. And Quine stated in 1950, "The variables remain mere pronouns, for cross-reference; just as 'x' in its recurrences can usually be rendered 'it' in verbal translations, so the distinctive variables 'x', 'y', 'z', etc., correspond to the distinctive pronouns 'former' and 'latter', or 'first', 'second', and 'third', etc." [4].

The thesis of this article is to suggest that the logicians' view of variables is best for the teaching of mathematics—that, right from the beginning and regardless of whether they are called "letters," "literals," "literal symbols," or "variables," they should be described as placeholders, and that, to be seen as meaningful, they should be presented in full sentences, especially ones with quantification. This thesis is supported by providing a sampling of the different uses of variables and analyzing the reasons for some of the difficulties students encounter with them. Two that arise repeatedly are (1) thinking of variables as exotic mathematical objects that do not have a clear connection with our everyday universe, and (2) regarding variables as having an independent existence even though they have been introduced as bound by a quantifier.

Mathematical Uses of Variables

Variables Used to Express Unknown Quantities

In the early grades, students are sometimes given problems like the following:

Find a number to place in the box so that $3 + \square = 10$.

Later, however, when algebra is introduced, the empty-box notation is typically abandoned and the focus shifts to learning rules for manipulating equations to get a variable, typically x, on one side and a number on the other. With the resulting emphasis on mechanical procedures, the meaning of "Solve the equation for x" may be obscured, with students coming to view x as a mysterious object with no relation to the world as they know it. Pointing out that x just holds the place for the unknown quantity—perhaps even making occasional use of the empty-box notation even after variables have been introduced—can counteract students' sense that the meaning of x is beyond their understanding.

To solve an equation for x simply means to find all numbers (if any) that can be substituted in place of x so that the left-hand side of the equation will be equal to the right-hand side. In my work with high school mathematics teachers, I have found that a surprising number are unfamiliar with this way of thinking and have never thought of asking their students to test the truth of an equation for a particular value of the variable by substituting the value into the left-hand side and into the right-hand side to see if the results are equal.

By holding the place for the unknown quantity in an equation such as $\sqrt{4 - 3x} = x$, the variable x enables us to work with it in the same way that we would work with a number whose value we know, and this is what enables us to deduce what its value or values might be. In 1972, the mathematician Jean Dieudonné characterized this approach by writing that when we solve an equation, we operate with "the unknown (or unknowns) as if it were a known quantity. . . . A modern mathematician is so used to this kind of reasoning that his boldness is now barely perceptible to him" [1].

VARIABLES USED IN FUNCTIONAL RELATIONSHIPS

Understanding the use of variables in the definition of functions is critically important for students hoping to carry their study of mathematics to an advanced level. In casual conversation, we might say that as we drive along a route, our distance varies constantly with the time we have traveled. So if we let d represent distance and t represent time, it may seem natural to describe the relationship between t and d by saying that for each change in t there is a corresponding change in d. This language has led many to think of variables such as t and d as objects with the capacity to change. Indeed, the word variable itself suggests such a description.

Addressing this issue, however, Tarski wrote, "As opposed to the constants, the variables do not possess any meaning by themselves. . . . The 'variable number' x could not possibly have any specified property . . . the properties of such a number would change from case to case . . . entities of such a kind we do not find in our world at all; their existence would contradict the fundamental laws of thought" [5]. Quine expressed a similar caution: "Care must be taken, however, to divorce this traditional word of mathematics [variable] from its archaic connotations. The variable is not best thought of as somehow varying through time, and causing the sentence in which it occurs to vary with it" [4].

We are quick to correct students who write, "Let a be apples and p be pears," telling them that they should say "Let a be the number of apples and p be the number of pears." Similarly, t does not actually represent time but holds a place for substituting the number of hours we have been driving, and d does not actually represent distance but holds a place for substituting the corresponding number of miles traveled during that time. Thus, it is not the t or the d that changes; it is the values (number of hours or number of miles) that may be put in their places. However, this is a distinction that mathematics teachers rarely emphasize to their students. In fact, mathematicians frequently make statements such as, "As x gets closer and closer to 0, $1 - x$ gets closer and closer to 1." This way of describing a variable that represents a numerical quantity may contribute to students' common belief that the number $0.99999 \ldots$ "gets closer and closer to 1 but it never reaches 1."

Even more than in the other areas of mathematics they encounter, students must learn to translate the words we use when we describe a function into language that is meaningful to them. For example, we might refer to "the function $y = 2x + 1$." Taken by itself, however, "$y = 2x + 1$" is meaningless. It is simply a predicate, or open sentence, that only achieves meaning when particular numbers are substituted in place of the variables or when it is part of a longer sentence that includes words such as "for all" or "there exists."

Students need to learn that when we write "the function $y = 2x + 1$," we mean "the relationship or mapping defined by corresponding to any given real number the real number obtained by multiplying the given number by 2 and adding 1 to the result." We think of x as holding the place for the number that we start with and y as holding the place for the number that we end up with, and we call x the "independent variable" because we are free to start with any real number whatsoever and y the "dependent variable" because its value depends on the value we start with. Imagining a process of placing successive values into the independent variable and computing the corresponding values to place into the dependent variable can give students a feeling for the dynamism of a functional relationship. However, we need to alert students to the fact that the specific letters used to hold the places for the variables have no meaning in themselves. For example, the given function could just as well be described as "$v = 2u + 1$" or "$q = 2p + 1$," or as "$x \rightarrow 2x + 1$" or "$u \rightarrow 2u + 1$."

Another way to describe this function is to call it "the function $f(x) = 2x + 1$" or, more precisely, "the function f defined by $f(x) = 2x + 1$ for all real numbers x." An advantage of the latter notation is that it leads us to think of the function as an object to which we are currently giving the name f. This notation also makes it natural for us to define "the value of the function f at x" as the number that f associates to the number that is put in place of x. Using the notation $f(x)$ to represent both the function and the value of the function at x, while convenient for certain calculus computations, can be confusing to students.

A variation of the preceding notation defines the function by writing $f(\square) = 2 \cdot \square + 1$, pointing out that for any real number one might put into the box, the value of the function is twice that number plus 1. The empty box representation is especially helpful for work with composite functions. Students asked to find, say, $f(g(x))$ often become confused when both f and g have been defined by formulas that use x as the independent variable. When the functions have been defined using empty boxes the relationships are clearer. For instance, in a calculus class students find it easier to learn to compute $f(x + h)$ if they have previously been shown the definition of f using empty boxes.

VARIABLES USED TO EXPRESS UNIVERSAL STATEMENTS

Terms like "for all" and "for some" are called quantifiers because "all" and "some" indicate quantity. In a statement starting "For all x" or "For some x," the "scope of the quantifier" indicates how far into the statement the role played by the variable stays the same, and the variable x is said to be "bound" by the quantifier.

Most mathematical definitions, axioms, and theorems are examples of universal statements, i.e., statements that can be written so as to start with the words "for all." For example, the distributive property for real numbers states that for all real numbers a, b, and c, $ab + ac = a(b + c)$. The variables a, b, and c are bound by the quantifier "for all," and they are placeholders in the sense that no matter what numbers are substituted in their place, the two sides of the equation will be equal. Thus the symbols used to name them are unimportant as long as they are consistent with the original.

In mathematics classes it is common to abbreviate the distributive property (and similar statements) by saying that a certain step of a solution is justified "because $ab + ac = a(b + c)$." However, this usage can lead students to invest a, b, and c with meaning they do not actually have. For instance, some students become confused when asked to apply the distributive property to $cb + ca$ because the a, b, and c are the same symbols used in the statement of the property, and students think of them as continuing to have the same meaning as in the statement, without realizing that the scope of the quantifier extends only to the statement's end.

A different problem arises when the omission of the quantifiers is justified by describing a, b, and c as "general numbers" because this name suggests that there is a category of number that lies beyond the ordinary numbers with which students are familiar. For those with a secure sense of the way a, b, and c function as placeholders, this terminology is not misleading, but students with a shakier sense of the meaning of "variable" may imagine a realm of mysterious mathematical objects whose existence makes them uneasy.

By contrast, if the distributive property is simply described as a template into which any real numbers (or expressions with real number values) may be placed to make a true statement, the mystery disappears and the way is prepared for leading students to an increasingly sophisticated ability to apply the property. Again empty boxes may be helpful. For example, the property can be stated as follows: No matter what real numbers we place in boxes $\square$, $\lozenge$, and $\triangle$,

$$\square \cdot \lozenge + \square \cdot \triangle = \square \cdot (\lozenge + \triangle)$$

Encouraging students to test the template by substituting a variety of different quantities in place of $\square$, $\lozenge$, and $\triangle$ provides a gentle introduction both to the logical principle of universal instantiation[8] and to the dynamic aspect of the universal quantifier, and substituting successively more complication expressions into the boxes can develop a sense for the power of the property:

$$2 \cdot s + 2 \cdot t = 2 \cdot (s + t)$$
$$2s + 6 = 2 \cdot s + 2 \cdot 3 = 2 \cdot (s + 3)$$
$$2^{100} + 2^{99} = 2^{99} \cdot 2 + 2^{99} \cdot 1 = 2^{99} \cdot (2 + 1) \, [= 2^{99} \cdot 3]$$
$$(x^2 - 1) \cdot x + (x^2 - 1) \cdot (x - 3) = (x^2 - 1) \cdot (x + (x - 3)) \, [= (x^2 - 1)(2x - 3)]$$

DUMMY VARIABLES AND QUESTIONS OF SCOPE

Strictly speaking, the term dummy variable simply refers to any variable bound by a quantifier, but we most often use the term when discussing summations and integrals. For example, given a sequence of real numbers $a_0, a_1, a_2, \ldots$ and a function f, we make a point of referring to k, i, x, and t as dummy variables to help students understand that

$$\sum_{k=1}^{10} a_k = \sum_{i=1}^{10} a_i \quad \text{and} \quad \int_1^2 f(x)dx = \int_1^2 f(t)dt.$$

In fact, it may be helpful to use the term dummy variable whenever we are especially concerned about problems that can result from thinking of variable names as "exceeding their bounds," that is, as having meaning outside the scope determined by their quantification. For instance, it is common to state the definitions of even and odd integers as follows:

> For an integer to be even means that it equals $2k$ for some integer k.
> For an integer to be odd means that it equals $2k + 1$ for some integer k.

Following such an introduction, many students try to prove that the sum of any even integer and any odd integer is odd by starting their argument as follows:

> Suppose m is any even integer and n is any odd integer. Then $m = 2k$ and $n = 2k + 1 \ldots$

For the definitions of even and odd, however, the binding of each occurrence of k extends only to the end of the definition that contains it. To avoid the mistake shown in the example, students must come to understand that the symbol k is just a placeholder, with no independent existence of its own. One way to emphasize this fact is to call k a dummy variable. We can reinforce this characterization by writing each definition several times, using a different symbol for the variable each time. For example. we could write the definition of even as:

> For an integer to be even means that it equals $2a$ for some integer a.
> For an integer to be even means that it equals $2r$ for some integer r.
> For an integer to be even means that it equals $2m$ for some integer m.

It is also effective to give an alternative version of the definition that does not use a variable at all:

> For an integer to be even means that it equals twice some integer.

In general, asking students to translate between formal statements that contain quantifiers and variables and equivalent informal statements without them is helpful in developing their ability to work with mathematical ideas.

A few years ago, I discovered that when I asked students to write how to read, say, the following expression out loud:

$$\{x \in U \mid x \in A \text{ or } x \in B\}$$

the most common response was to omit the words "the set of all" and write only "x in U such that x is in A or x is in B." More recently, when teaching about equivalence relations, I learned that part of students' difficulty in interpreting such a set definition was a belief that the variable x had a life outside of the set brackets. When I defined the equivalence class of an element a for an equivalence relation R on a set A as

$$[a] = \{x \in A \mid x \, R \, a\},$$

a number of students had trouble applying the definition, and the question they asked was, "What happened to the x?" However, they were successful after I showed them that the definition could be rewritten with t in place of x and that it could be rephrased without the x as "The equivalence class of a is the set of all elements in A that are related to a."

Instructors who teach students with computer programming experience can draw analogies between the ways variables are used in programs and the ways they are used in mathematics. For example, the name for a "local" variable in a subroutine can be used with a different meaning outside the subroutine, and within the subroutine it can be replaced by any other name as long as the replacement is carried out consistently. This is strikingly similar to the way a mathematical variable acts within a definition or theorem statement.

Variables Used as Generic Elements in Discussions

A variable is sometimes described as a mathematical "John Doe" in the sense that it is a particular object that shares all the characteristics of every other object of its type but has no additional properties. For example, if we were asked to prove that the square of any odd integer is odd, we might start by saying, "Suppose n is any odd integer." As long as we deduce properties of n^2 without making any assumptions about n other than those satisfied by every odd integer, each statement we make about it will apply equally well to all odd integers. In other words, we could replace n by any odd integer whatsoever, and the entire sequence of deductions about n would lead to a true conclusion. In that sense, n is a placeholder.

To be specific, consider that, by definition, for an integer to be odd means that it equals 2 times some integer plus 1. Because this definition applies to every odd integer, a proof might proceed as follows:

Proof: Suppose n is any odd integer. By definition of odd, there is some integer m so that $n = 2m + 1$. It follows that

$$n^2 = (2m + 1)^2 = 4m^2 + 4m + 1 = 2(2m^2 + 2m) + 1.$$

But $2m^2 + 2m$ is an integer, and so n^2 is also equal to 2 times some integer plus 1. Hence, n^2 is odd.

Dieudonné's use of the word "boldness" to describe the process of solving an equation by operating on the variable as if it were a known quantity applies equally well to the use of a variable as a generic element in a proof. For instance, by boldly giving the name n to an arbitrarily chosen, but representative, odd integer, we can investigate its properties as if we knew what it was. Then, after we have used the definition of odd to deduce that n equals two times some integer plus 1, we can boldly apply the logical principle of existential instantiation[2] to give that "some integer" the name m in order to work with it also as if we knew what it was.

Occasionally we may be given a problem in a way that asks us to think of a certain variable as generic right from the start. For instance, instead of being asked to prove that the square of any odd integer is odd, we might have been given the problem: "Suppose n is any odd integer. Prove that n^2 is odd." In this case, after reading the first sentence, we should think of n as capable of being replaced by any arbitrary odd integer, and we would omit the first sentence of the proof that is given above.

An important use of variables as generic elements in mathematics education occurs in deriving the equations of lines, circles, and other conic sections. For example, to derive the equation of the line through $(3, 1)$ with slope 2, we could start as follows: "Suppose (x, y) is any point on the line." As long as we deduce properties of x and y without making any additional assumptions about their values, everything we conclude about (x, y) will be true no matter what point on the line might be substituted in its place.

We could continue by considering two cases: the first in which $(x, y) \neq (3, 1)$ and the second in which $(x, y) = (3, 1)$. For the first case, we note that what ensures the straightness of a straight line is the fact that its slope is the same no matter what two points are used to compute it. Therefore, if the slope is computed using (x, y) and $(3, 1)$, the result must equal 2:

$$\frac{y - 1}{x - 3} = 2, \quad \text{and so} \quad y - 1 = 2(x - 3). \, (*) \tag{1}$$

This concludes the discussion of the first case. In the second case, $(x, y) = (3, 1)$ and both sides of Equation (1) equal zero. So in this case, it is also true that $y - 1 = 2(x - 3)$. Therefore, because no assumptions about (x, y) were made except for its being a point on the line, we can conclude that every point (x, y) on the line satisfies the equation $y - 1 = 2(x = 3)$.

Conclusion

This chapter has advocated placing greater emphasis on the role of variables as placeholders to help address students' difficulties as they make the transition to algebra and more advanced mathematical subjects. Supporting examples were given from a variety of mathematical perspectives. It is hoped that the paper will stimulate additional research to delve more deeply into the issues it raises.

Notes

1. Universal instantiation: If a property is true for all elements of a set, then it is true for each individual element of the set.

2. Existential instantiation: If we know or suspect that an object exists, then we may give it a name, as long as we are not using the name for another object in our current discussion.

References

1. Dieudonné, J., "Abstraction in mathematics and the evolution of algebra." In: Lamon, W. (ed.) *Learning and the Nature of Mathematics*, SRA Associates, Inc., Chicago (1972), pp. 102–103.

2. Frege, G., *The Basic Laws of Arithmetic: Exposition of the System* (Translated and edited by M. Furth), University of California Press, Berkeley, CA (1964).

3. Klein, F., *Elementary Mathematics from an Advanced Standpoint* (Translated from the 3rd German edition by Hedrick, E. R., and Noble, C. A.), Dover Publications, New York (1924).

4. Quine, W.V.O., *Methods of Logic*, 4th ed. Harvard University Press, Cambridge, MA (1982).

5. Tarski, A., *Intrtoduction to Logic and to the Methodology of Deductive Sciences*. Oxford University Press, New York (1941).

6. Whitehead, A. N., *An Introduction to Mathematics*, Henry Holt and Co., New York (1911).

Bottom Line on Mathematics Education

DAVID MUMFORD AND SOL GARFUNKEL

First some axioms: Mathematics is honestly useful for all citizens. It can help them in school, at work, as citizens and in their daily lives. This is the reason we teach mathematics every year from kindergarten through the end of high school. The mathematical education of the general public is a priority of our educational system above and beyond the education of future mathematicians and scientists.

What follows from these axioms is that we need a system of mathematics education that seeks first and foremost to recognize the mathematical needs of average citizens and is designed to ensure that those needs are met, while hopefully meeting the needs of students who can and want to learn mathematics as a discipline. And we need to acknowledge that for quite a while now we have been doing precisely the reverse. In other words, we have designed a system for the mathematically motivated and talented and let others drop away—without regard to whether they would be able to use the fragmentary mathematical understandings with which they were left.

The latest attempt at mathematics education reform, the Common Core State Standards in Mathematics (CCSSM), exemplify this failed approach. Will it help us do better as a country on international comparisons? Possibly. Will it create a more balanced playing field across the country? Very likely. Will it make it easier to identify talented mathematics students earlier in their school careers? Almost certainly. Will it move us toward a more quantitatively literate population and work force? Absolutely NOT!

The CCSSM are being marketed as college and career-readiness standards, with the implication that they are for everyone. This is falsity in advertising. If we want future adults to learn to use mathematics, then we must show them how mathematics is used in ways and situations that are genuine and that are relevant to their own experience. This isn't really all that hard. The truth is that most mathematics was invented to solve practical and interesting problems. Rather than spend year after year learning more and more abstract and sophisticated tools, we can take some of that time to use those tools to build real things. Mathematics is a system that enables us to model the world. We need to let students in on this fact—to actually have them use the mathematics they are learning to do what it was meant to do—give them a greater ability to understand the world around them.

The standards and the high-stakes tests that are being developed (from 3rd grade on!!) certainly make for more consistency from state to state. But consistency should not be a goal in itself. The mathematical literacy of the next generation is a goal we should be working toward. And to achieve that goal, we need a true reorganization of the mathematics that we teach—keeping why we teach it prominently up front for all to see.

What Might This Brave New World Look Like?

- For one thing, it's obvious that everyone's lives revolve around money, and certain types of math are an essential in mastering fiscal challenges. It is not hard to compare the real cost of leasing vs. buying a car if you know the right math but nearly impossible without it. Problems in school based on analyzing the actual budgets of people, businesses, and countries could improve our pathetically math-impoverished political discourse.
- Everyone says computer technology should be used in schools, but why let the computer be another incomprehensible technological mystery? Teach everyone the rudiments of programming and what goes on inside that box. "But is this math?" we hear you saying. Yes; writing computer code teaches you how to be precise and formal and makes concrete mathematical recipes like that for long division. They are what we call algorithms, and this sort of training is a paradigm for rational thinking.
- Data need not be something we leave to white-coated experts. Students can get tons of data on prices from newspapers and go to town with means and standard deviations. We would suggest asking students to tally their caloric intakes and find

the correlation between this and their weights—but maybe some math is too painful to see. Nonetheless, adults struggle to absorb medical recommendations based on similar statistics ("lies, damned lies and statistics," as a well-known figure said), and arguing about the validity of statistics can't start too early.

- One could go on. The key lesson is that so much of what is going on in the world can be modeled mathematically. Maps, music, the range of a rifle, how to rig elections, the sustainability of a fishery—you name it!

The CCSSM do try to include some applications like this. But instead of building theory step by small step out of easily absorbed and useful applications, they tag on an application or two as an afterthought to a heavy slog through abstraction. Math professors know well how to explain math to graduate students, but they seem to have forgotten how high they have climbed into the clouds and what is going on down here on the Earth.

History of Mathematics and History of Science Reunited?

JEREMY GRAY

How to write the history of modern mathematics? This question, in itself no harder or less capable of an answer than the broader question of how to write the history of modern science, should be part of that broader question, but it has become separated. Recent initiatives, however, suggest that these questions can once again be raised and discussed together. There are in each case several fundamental latent issues. The question of how we should do something invites us to consider who "we" are and for whom we are doing it. I duck the first of these considerations and note that different audiences want different things from the history of science and the history of mathematics, so this essay will necessarily have to make some uncomfortable compromises. The same is true of the words "modern" and "mathematics." I shall restrict my attention to the mathematics of the long nineteenth and short twentieth centuries and, further, to the activities of professional mathematicians because this is where the separation between historians of mathematics and of science has become greatest and where, perforce, some of the most innovative work has been done. The secret history that George Sarton referred to so many years ago is buried most deeply here and must be brought out.[1]

The audience question is itself multifaceted. The disciplinary divide works differently in mathematics and in science, and there is nothing like the separation in the development of mathematics that the theories of general relativity and quantum mechanics create in physics. Indeed, much of today's undergraduate mathematics syllabus is the creation of nineteenth-century mathematicians. So one audience for the history of modern mathematics, quite rightly, is professional mathematicians. Inevitably, these people have expectations that historians of science do not, and the compromises authors make when writing for an audience of mathematicians have disadvantaged them when writing for the second group. But this is a smaller matter than it has been allowed to appear. Every act of writing, every piece of research, is a series of compromises: with the sources, with one's own skills (linguistic, financial, and so forth), with one's knowledge of other fields (social, historical, political, philosophical—the list is only too long). The only way we have as historians of coping with all these demands is the collective nature of our work and the possibility of relying on the knowledge of others. If, from a certain perspective, all a lengthy technical history of a piece of modern mathematics does is to establish that a particular mathematician spoke with well-grounded authority, then that might be enough. Better that such a claim be established than that it be taken on trust, because in mathematics, as in the rest of science, authority is only partial, dynamic, and contested.

The challenge posed to historians of modern mathematics by the contemporary history of science is to move away from a worn-out mode of history of ideas, a challenge exacerbated by the highly specialized nature of the ideas themselves. This challenge exists in several forms. There are those who see big themes in the history of science, such as mechanical objectivity, and there are advocates of highly localized studies (one site over a fairly short period of time). There are those who would finesse the difficult technical material and those (perhaps a smaller number these days) who would savor its variety. There are those who would have us see "real people" in all their historical contingency. But perhaps the dominant pressure is to move the history of science away from an intense focus on scientific ideas and on to the integration of science in society: its uses, its costs, its political implications. None of this is easy. It is not impossible, for example, to write a biography of a scientist that integrates the person and his or her times, but the very fact that it is a commonplace in the criticism of biographies of poets and novelists that the links between "life" and "works" are less than one would like can surely stand as a sign that it will be much harder to anchor a modern mathematician in his or her context and, still more, that it will be difficult to "explain" much about him or her. See Joan Richards's account of how this can be done in the happy case of Augustus De Morgan.[2]

The hardest task for the historian of mathematics is surely to keep the mathematical ideas in play, and I shall now restrict this essay to a consideration of that task. Much effort has recently gone into finding ways of talking about significant aspects of modern mathematics that do not presume that it developed on what Theodore Porter recently called an "island of technicality."[3] Interestingly, these efforts often run parallel to a new movement among philosophers to write about mathematical practice, and the parallels are worth spelling out. Nonetheless, this new development is paradoxical because the most sterile way to see mathematics is as a body of statements validly deduced from some set of initial assumptions or axioms, and this view of mathematics has been that of philosophers of mathematics for some time. It is correct as far as it goes—but that is not very far, and it ignores many of the things that mathematicians themselves find important in mathematics. Mathematics is done for a variety of reasons and in a variety of ways that mathematicians sometimes argue about. It is done in a variety of contexts and settings. It is a dynamic enterprise, with evolving objectives and a curious coherence. A view of mathematics that fits only poorly to the actors' views, that fits only poorly to current conceptions of knowledge, and that does not allow for the creation of new knowledge is surely not one any historian can have.

There were, however, good reasons for philosophers of mathematics to accept it. The founding fathers of modern mathematical logic, people like Kurt Gödel and Alfred Tarski, took up issues in the foundations of mathematics, such as the nature of proof and truth, and created a magnificent tool for addressing these questions—one that resolved many of them, sometimes in surprising ways. It was natural for their intellectual descendants to adopt a reductionist position according to which all of mathematics reduces to set theory in its various forms and all of its philosophical questions can be concentrated into questions about axioms and their consequences.

It was also natural for these philosophers to follow philosophers of science in dismissing many other questions about mathematics, such as how new mathematics can be discovered, as psychological and outside the business of the philosopher. This type of philosophy of mathematics seldom appeals to mathematicians, and more recently some of the best mathematical logicians have swung their subject around to make it produce results of greater interest to working mathematicians. As a by-product of this swing, a number of philosophers of mathematics have begun to ask philosophical questions about mathematics that are more in tune with good questions for historians to ask and that, moreover, are also asked by practicing mathematicians, as is demonstrated by the essays in two recent books edited by Paolo Mancosu and by José Ferreirós and me. We can also note that the highly conceptual character of modern mathematics has produced a correspondingly sophisticated response in Moritz Epple's work on epistemic objects.[4] Leo Corry's useful distinction between the body and the image of mathematics is another promising way of shifting attention to important questions in the history of modern mathematics that unite the intellectual and the social, contextual aspects of mathematics.[5]

Some of these questions are entirely traditional: How is mathematics discovered? What are the roles of heuristic arguments, conjectures, open problems, big programs, and plans of research? Is mathematics about problem solving or theory building? Some questions are newer: What do mathematicians mean when they insist on purity of method or the unity of mathematics? What is it for a concept to be "fruitful" or "natural"? What is it for a mathematician to complain of a valid proof that it is nonetheless the "wrong" proof? Some questions are about important but specific points: Why do mathematicians often seek new proofs of known results? What is a theorem, other than the conclusion of a proof? What is a good theorem? What roles do theorems play in mathematics? What is an answer to a mathematical question? All these questions have their obvious analogous historical formulations, and the grist that historians provide to the new philosophers' mills is returned in the form of ways of thinking about mathematics that offer ways to say significant things about the creation of modern mathematics.

These questions look quite different when mathematics is taken in relative isolation from its applications in the hard sciences and when they are taken together. This being the case, the historian naturally turns to examine this apparent split. Is it a real distinction or an artifact (an "elephant" question)? If there is anything substantial to the distinction, how did it come about? Does it vary from place to place? How should historians deal with it? One view that has been presented at length in the past decade or so is that there is a valid, culturally and socially grounded sense in which one can talk of a distinctively pure mathematics, and its emergence can be regarded as the emergence of a modernism akin to the better-known modernisms in painting, music, fiction, and the other arts of the early twentieth century. The pioneer here was Herbert Mehrtens, who argued that especially in Germany there emerged a modernist (that is, abstract) mathematics that was in fruitful, dialectical opposition to a countermodern movement (more intuitive and more oriented toward applications). His book *Moderne Sprache Mathematik*, which deserves to be much better known but has never been translated out of German, drew a generally accepted picture of these two camps, albeit in a largely German setting, before engaging with broader questions about the nature of modernity and, always in the author's mind, the rise of Nazi Germany. William Everdell's *The First Moderns*, which is culturally more wide ranging, should also be noted for placing discontinuity in mathematics at the head of a panoply of changes that ushered in modernism.[6]

For the claim that there was a genuine modernist mathematics to make any sense, there must be at least a rough-and-ready definition of modernism that fits the artistic cases and against which the mathematical case can be measured (and found to succeed or fail). In *Plato's Ghost*, I offered this capsule definition: Modernism is an autonomous body of ideas, having little or no outward reference, placing considerable emphasis on formal aspects of the work, and maintaining a complicated— indeed, anxious—rather than a naive relationship with the day-to-day world, that is the *de facto* view of a coherent group of people, such as a professional or discipline-based group, who have a high sense of the seriousness and value of what they are trying to achieve.[7]

I then argued that there was a movement toward a new kind of mathematics, variously called "abstract," "free," or "pure" mathematics by its practitioners, that by 1900 came to be the dominant form of research mathematics in Germany—the leading center for mathematics in the world at the time—and that made characteristic changes to many branches of mathematics: in geometry, analysis, algebra, and the new subject of mathematical logic.

In each of these areas, the new "modern" mathematics was done, one could say, for its own sake. Definitions were internal to the subject and no longer derived by abstraction from real-world objects, even in geometry. Mathematical objects had merely to obey the rules of the mathematician and could be profitably studied without any hope or intention of application. First in German universities, and then in Italian and French ones, but to a significantly lesser extent in Britain, this new mathematics was the one that was taken to capture the essence of the subject. It is in this setting that questions of purity of method and the unity of mathematics were most forcefully raised, where mathematicians debated in public about the nature of proof and the meaning of mathematics, looked for the "right" explanation and not just any valid explanation, and publicly established the intellectual and institutional independence of the mathematical enterprise.

Even if we grant, for a moment, the validity of the claim that there was a modernist phase in mathematics, we must ask: So what? What is in this sweeping picture that we did not see so clearly before? It gives a fresh picture of the rise of "pure" mathematics, one that makes historically specific the tension that has existed ever since between mathematics and its applications (and may have contributed, in passing, to the separation of the history of mathematics from the history of science with which this chapter is concerned). It allows the historian to look anew at the relation of modern mathematics to the sciences, and it directs attention to the wholesale change in the philosophy of mathematics associated with the names of Frege, Hilbert, and others that the emergence of modern pure mathematics provoked, as the new philosophy became the ideology of the new mathematics. Talk of axiom systems, syntax, and semantics became more prominent, and mathematics was spoken of as a formal "language," which invites the historian to consider contemporary developments in the study of linguistics. Personally, I enjoyed the idea that the history of music might become available as a further model for the historian of mathematics. But what has struck a chord with at least one mathematician reviewing *Plato's Ghost* was the theme of anxiety. This is familiar in all writing on the artistic modernisms, especially in the setting of discussions of modernity, and it can be traced in the emergence of mathematical modernism too. As each new difficulty in emancipating mathematics from its pragmatic origins emerged, new, deeper escape routes were offered, until by 1900 the grounds for the acceptance of mathematics at all were in question. Famously, the attempt (by Frege, Russell, Whitehead, and others) to derive mathematics from logic failed, later attempts to derive it from axiomatic set theory were found forever insufficient by the work of Gödel, and radical alternatives such as Brouwer's intuitionism came up short as well. There was to be no escape, no certainty, even in mathematics.

What big claims of this kind do for historians, if they prove lasting, is to set the scene for other debates. To speak of "a" or "the" scientific revolution, a republic of letters, the mechanical worldview, Enlighten-ment science, or mechanical objectivity is to invite criticism. Does this sweeping claim hold up, and even if so does it encompass everything and therefore say nothing? And if it does, can it be made usefully more particular by comparing one locale with another, or one smaller period with another? Mathematical modernism as a phenomenon active be-tween 1880 and 1920, with some outlying figures and some wrinkles, may hold up as an account of the rise of pure mathematics for historians already well disposed to grant a distinction between pure and applied mathematics, such as the one on display in most mathematics depart-ments in most of the twentieth century. The sharper test will come from those more inclined to see mathematics as inseparable from its applications and uses, even if the marriage of mathematics and modern physics is seen as a rocky one.

The most valuable contribution to this debate in recent years has been Leo Corry's *David Hilbert and the Axiomatization of Physics (1898–1918)*, which showed just how preoccupied Hilbert, the archmodernist in everyone's book, was with contemporary physics, on which he lec-tured almost every year between 1900 and 1918. Corry also found new documents illuminating the distinction between Hilbert's and Einstein's ideas about general relativity in the crucial year 1916.[8] The reformula-tions of special relativity in the interface between mathematicians and physicists have been carefully analyzed by Scott Walter; Erhard Scholz has written extensively on the work of Hermann Weyl (from physics to philosophy)—as have Skuli Sigurdsson *et al.*[9] And there is every rea-son to suspect that the comprehensive survey of the reception of Ein-stein's general theory that Jim Ritter is working on will further change the picture we have of this deep intermingling of applied mathematics and theoretical physics. Despite all the work that has been done on the early history of quantum mechanics, we have until recently lacked the view from the history of mathematics, but we shall soon have Martina Schneider's comparative study of the work of Bartel Van der Waerden, John von Neumann, and Eugene Wigner.[10]

Of these analyses, most have a local flavor, heavily Göttingen cen-tered or at least Göttingen influenced (only Ritter is ranging widely). There are other locally based investigations—for example, by Epple and his students into the Vienna situation. There are also investigations into early mathematical aeronautics (the theme of a special session at the 23rd International Congress in the History of Science and Technology, held in Budapest in 2009) and, more specifically, a treatment of the life and work of Richard von Mises by Reinhard Siegmund-Schultze that will explore the transformation of applied mathematics around the time of World War I. The effect of the war on mathematics is also the subject of a forthcoming volume edited by David Aubin and Catherine Goldstein. By no means all of these studies concern themselves with the modernist claim, but they do bring into historical focus for the first time some of the most important ways mathematics has been involved in the creation of the modern world.

The history of modern science, and within it the history of modern mathematics, can, of course, never be written; it is simply too large a subject. But, like the elephant, parts of it can be written about—and written about in different ways. There is no reason to force even the overlapping parts into a strict agreement: critical standards are compatible with conflicting perspectives. In this essay I have identified two specific areas where there is a new history of mathematics that historians of science can respond to. The claim that there was a characteristically modernist phase in the development of mathematics is one area that suggests the possibility of links to other intellectual domains. The revived history of applied mathematics and theoretical physics is the other.

It is always possible to see other works as boundaries, and much computer ink has been spilled in recent years on the subject of boundaries (Are they barriers or are they permeable "trading zones"?). Historians of mathematics have of course read *How Experiments End*, *The Intellectual Mastery of Nature*, the outpourings of the Einstein industry, *Masters of Theory, Energy and Empire*, and other major works in the history of science that overlap with their own field.[11] These richly suggestive works have left much for historians of mathematics to do—for example, we still lack a mathematically adequate account of how a Cambridge education in mathematics in the nineteenth century led to such a powerful school of applied mathematicians in Britain—but a better opportunity now exists for cooperative and interactive work in the history of mathematics and the history of science than we have had for 20 years. After all, mathematics of all kinds has been an intimate part of physics for centuries and is increasingly involved in all of today's new sciences. There is no reason why the relations between the history of mathematics and the history of science should not be as they are in mathematics and science themselves: contested, disputatious, affirmed and denied, emphasized and marginalized, but undeniably productive and mutually beneficial.

In this endeavor, the study of mathematical practice may prove to be helpful. Mathematicians do many more things than state and prove theorems. They evaluate the importance of results, and they use the best results to shape what they know and help direct their investigations. They may have a preference for one method over another (the question of purity of method) or accept anything that works; they may find some concepts fruitful; they may see and value a unity in mathematics. These and other important aspects of mathematical thinking raise new philosophical questions that can also guide the writing of the history of mathematics by focusing attention on significant aspects of mathematical research. The other side of this mirror is, of course, entirely social. These preferences, and how they are accepted, modified, or rejected, play out in careers, in the growth of schools and institutions, in the interactions with neighboring sciences, and in many other ways. The tentative character of mathematical research needs deeper study. Mathematicians, singly and in groups, commit themselves to topics, a complex of problems, and possible methods for tackling them in advance of any guarantee of success: The rise and fall of concepts goes along with the rise and fall of places and people. A theorem may stand for all time, as most mathematicians assert, but they also agree that it may not always stand at center stage. A problem, as Poincaré claimed, may only ever be more or less solved as insights into it yield both answers and deeper questions. Major topics for one generation may reshape into others or so defy challenge as to become marginal for lack of hope. A mathematical idea cannot be fully understood without a consideration of the techniques that make it accessible, the reasons advanced for its importance, the problems and topics with which it is, or might be, connected. All of this topic has a historical dimension that is palpable and dramatic, that sees mathematics as an activity (as mathematicians do) and locates it centrally in the view of historians. We have a lot to do.

Notes

1. George Sarton, *The Study of the History of Mathematics* (Harvard Univ. Press, Cambridge, MA, 1937; reprint, Dover, New York, 1957).

2. Joan Richards, "'This Compendious Language': Mathematics in the Work of Augustus de Morgan." Isis, 102.3, 2011, 506–510. A number of other integrated biographies of mathematicians could be mentioned. Among the Victorian British, e.g., see Tony Crilly, *Arthur Cayley: Mathematician Laureate of the Victorian Age* (Johns Hopkins Univ. Press, Baltimore 2006); and Karen Hunger Parshall, *James Joseph Sylvester: Jewish Mathematician in a Victorian World* (Johns Hopkins Univ. Press, Baltimore 2006).

3. Theodore M. Porter, "How Science Became Technical." Isis, 2009, 100, 292–309, on p. 309.

4. Moritz Epple, "Between Timelessness and Historiality: On the Dynamics of the Epistemic Objects of Mathematics." Isis, 102.3, 2011, 481–493.

5. Paolo Mancosu, ed., *The Philosophy of Mathematical Practice* (Oxford Univ. Press, Oxford, U.K. 2008); José Ferreirós and Jeremy Gray, eds., *The Architecture of Modern Mathematics* (Oxford Univ. Press, Oxford, U.K. 2006); and Leo Corry, *Modern Algebra and the Rise of Mathematical Structures*, 2nd ed. (Birkhauser, Boston/Basel 2003).

6. Herbert Mehrtens, *Moderne Sprache Mathematik: Eine Geschichte des Streits um die Grundlagen der Disziplin und des Subjekts Formaler Systeme* (Suhrkamp, Frankfurt am Main 1990); and William R. Everdell, *The First Moderns: Profiles in the Origins of Twentieth-Century Thought* (Univ. Chicago Press Chicago 1997).

7. Jeremy Gray, *Plato's Ghost: The Modernist Transformation of Mathematics* (Princeton Univ. Press, Princeton, NJ 2008).

8. Leo Corry, *David Hilbert and the Axiomatization of Physics (1898–1918)* (Archimede, 10) (Kluwer, Dordrecht 2004); and Corry, Jürgen Renn, and John Stachel, "Belated Decision in the Hilbert–Einstein Priority Dispute." Science, 1997, 278, 1270–1273.

9. For Walter's analysis, see Scott Walter, "Breaking in the 4-Vectors: The Four-Dimensional Movement in Gravitation, 1905–1910." in *The Genesis of General Relativity*, Vol. 3: *Gravitation in the Twilight of Classical Physics: Between Mechanics, Field Theory, and Astronomy*, edited by Jürgen Renn and Matthias Schemmel (Springer, Berlin 2007), pp. 193–252. For Scholz's work on Weyl, see, e.g., Erhard Scholz, "Hermann Weyl's Analysis of the 'Problem of Space' and the Origin of Gauge Structures," Science in Context, 2004, 17, 165–197; and the works cited therein. See also Skuli Sigurdsson, Scholz, Hubert Goenner, and Norbert Straumann, "Historical Aspects of Weyl's *Raum Zeit Materie*," in *Hermann Weyl's* Raum—Zeit—Materie *and a General Introduction to His Scientific Work*, ed. Scholz (DMV Seminar, 30) (Birkhäuser, Basel 2001).

10. Schneider, M. *Zwischen zwei Disziplinen, B. L. van der Waerden und die Entwicklung der Quantenmechanik.* (Springer, Berlin, Heidelberg 2011).

11. Peter Galison, *How Experiments End* (Univ. Chicago Press, Chicago 1987); Christa Jungnickel and Russell McCormmach, *The Intellectual Mastery of Nature: Theoretical Physics from Ohm to Einstein* (Univ. Chicago Press, Chicago 1986); Andrew Warwick, *Masters of Theory: Cambridge and the Rise of Mathematical Physics* (Univ. Chicago Press, Chicago 2003); and Crosbie Smith and M. Norton Wise, *Energy and Empire: A Biographical Study of Lord Kelvin* (Cambridge Univ. Press, New York 1989).

Augustus De Morgan behind the Scenes

CHARLOTTE SIMMONS

"[I]t will ever be much more pleasing to grant even more praise
than is actually due, than to pluck the laurel
from the deserving brow."
—Benjamin Gompertz [15]

Augustus De Morgan (1806–1871) was a nineteenth century mathematician and prolific writer, author of more than 160 papers and 18 textbooks on algebra, arithmetic, trigonometry, probability, logic, and calculus, plus 850 articles in the popular, working-class oriented *Penny Cyclopedia* [7]. Here, however, we explore his contributions from behind the scenes, as a mentor to other mathematicians. Both Sir William Rowan Hamilton and George Boole, for example, two of the greatest algebraists of the nineteenth century, were close friends of De Morgan. During the period in which they produced some of their greatest work, both were in regular correspondence with De Morgan, who provided them with encouragement and reassurance in both their professional and personal lives. It is doubtful that either would have attained the level of success that they ultimately achieved without his help. De Morgan also provided support for the actuarial pioneer Benjamin Gompertz and the Indian mathematician Ramchundra, who has been called "De Morgan's Ramanujan" [18]. If De Morgan had done nothing else noteworthy in mathematics besides supporting the efforts of these four men and championing their work, we would still owe him a great debt.

Who Was De Morgan?

De Morgan entered Trinity College, Cambridge, at age 16, where he developed a lifelong love of mathematics. He obtained his bachelor's degree in 1827 and was offered the first Professorship of Mathematics at University College London the following year. Though he had no teaching experience, no publications, and was only 21, he was selected from 30 applicants "after the most distinguished competition that there has been for any chair," according to a letter from Thomas Coates [24]. The selection committee chose wisely. De Morgan became one of the longest serving and most highly respected professors at the university. Professor M. J. M. Hill (Astor Chair of Mathematics at London College from 1884 to 1924) said of him that "amongst the great men who have lectured within the walls of the College he was probably by reason of his scholarship, by the profundity of his work, and by his personal character, one of the greatest, if not the greatest of them all" [24]. By his retirement in 1867, De Morgan was one of the most distinguished mathematicians in Britain. In 1887, the London Mathematical Society established the De Morgan Medal in his honor.

De Morgan prided himself on being a champion of the underdog [23]. He is the only professor in the history of University College London to have resigned twice on matters of principle. His first resignation, in 1831, was in defense of a colleague whom he felt had been unfairly dismissed. De Morgan returned five years later, after his replacement drowned, to teach an additional 30 years. Then, when the most qualified candidate was denied a position at the college because he was a controversial Unitarian minister, De Morgan resigned again, never to return. He refused even to sit for a bust for the college library, explaining that as far as he was concerned his old college no longer existed [26].

Hamilton

The Irish mathematician Sir William Rowan Hamilton (1805–1865) is famous for constructing the first working algebraic system to break away from the real and complex numbers. Most importantly, Hamilton took the unheard-of step of abandoning the commutative law in formulating the quaternions (1843). In his *Lectures on Quaternions*, Hamilton credits De Morgan's attempts to extend the geometrical representation of complex numbers as having helped to lead him to this discovery.

Hamilton enjoyed a fast rise to the top. Born in Dublin at the stroke of midnight, "like Christ and Newton," by age 3 he had begun a rigid curriculum designed to prepare him for Trinity College [21]. At age 17, he was proclaimed a possible second Newton by Brinkley, Astronomer Royal of Ireland. He published his first paper while still an undergraduate; as a result of this work, he was unanimously elected to fill Brinkley's recently vacated position as Astronomer Royal and was appointed Professor of Astronomy at Trinity College just before completing his undergraduate degree. At age 30, he became the first person in Ireland ever to be knighted for scientific merit.

De Morgan and Hamilton

Though his reputation was established long before his correspondence with De Morgan began, De Morgan still provided valuable assistance to Hamilton. When Hamilton asked De Morgan to review his text, *Lectures on Quaternions*, he graciously agreed: "There is a pleasure in reading while any thing that strikes may do service" [13]. De Morgan never complained once, though the text grew to be 700 pages long and took five years to complete. De Morgan's task was thus no small undertaking. Though Hamilton expressed his "most sincere and unaffected conviction" that one who had had the benefit of a scientific education at a good university and attempted to study this text would find it "almost light reading," Herschel described the same work as one that would "take any man a twelvemonth to read, and near a lifetime to digest" [12].

As Hamilton once said, "[H]ow deeply man desires in intellectual things themselves the sympathy of man" [12]. De Morgan provided this sympathetic ear. Hamilton's "monstrous innovations" drew intense criticism at the time, but De Morgan provided support when Hamilton's quaternions were being attacked: "As to people ridiculing quaternions, let them do it; but do not let them succeed in making you feel it" [13].

Besides mathematics, their letters (only a portion fills 400 pages [13]) were filled with accounts of their children's activities, social events, poetry, politics, religion, and other details of their personal lives. Hamilton had this to say of their correspondence: "[H]e and I have exchanged a great number of pleasant letters, partly no doubt on mathematics, but we are not afraid to write nonsense to each other; at least I send him nonsense at times, and he sends me back wit in return, rising occasionally to humour" [13]. An example of this wit is the following, which De Morgan wrote to Hamilton after a lengthy break in their correspondence [13]:

If you are dead and buried, why do you not say so at once, like a man, instead of insinuating it in this roundabout way by solemn silence? What has become of the Manual of Quaternions? I write to you because I want to know something about you. . . . If you do not write I shall circulate a report that you have shipped yourself to fight for the Pope.

De Morgan also did his best to console Hamilton during times when "the earth seemed to him draped in black" [12]. When Catherine, the only woman he ever loved, married someone else, Hamilton contemplated suicide. He never really recovered emotionally and suffered from bouts of depression throughout his life. When she resurfaced years later, De Morgan received many letters detailing how the subject of Catherine "continued to agitate him to a degree beyond what is rational" [14].

Upon Hamilton's death, De Morgan wrote, "I have called him one of my dearest friends, and most truly; for I know now how much longer than twenty-five years we have been in intimate correspondence, of most friendly agreement or disagreement, of most cordial interest in each other. And yet we did not know each other's faces. I met him about 1830 at Babbage's breakfast table, and there for the only time in our lives we conversed. . . . And this is all I ever saw, so it has pleased God, all I shall see in this world of a man whose friendly communications were amongst my greatest social enjoyments, and greatest intellectual treats" [13].

Boole

George Boole (1815–1864) was almost entirely self-educated. Though one of his fellow pupils at the primary school he attended reports, "This George Boole was a sort of prodigy among us and we looked upon him as a star of the first magnitude," his parents were so poor that they could not afford to send him to secondary school, let alone college [16]. At 16, Boole had to seek employment as an assistant teacher to support his parents and siblings. It didn't go well. He was terminated for reading mathematics on Sunday and working math problems in chapel. Ultimately, he secured a professorship at a college in Ireland, where he taught until his death.

Boole's first major contribution to mathematics was the paper, "Exposition of a General Theory of Linear Transformations," which marked the beginning of algebraic invariant theory [2]; his next was "On a General Method in Analysis," which helped pave the way for operator theory [5]. But his greatest gift to mathematics was in logic. His first book on the subject, *The Mathematical Analysis of Logic* [4], is recognized as the beginning of modern symbolic logic. Here, Boole showed that logic is actually a branch of mathematics and introduced the notion that symbols need not always be interpreted quantitatively but can be used to represent objects as well. In addition, he developed a method for expressing the established rules of syllogistic reasoning algebraically, which others before him (including Leibniz) had attempted but failed.

De Morgan and Boole

Within mathematics, De Morgan is also best known for his work in logic. He created the first logic of relations and promoted a symbolic approach to the subject. Unfortunately, as De Morgan noted in his review of the work, Boole's masterpiece, *An Investigation of the Laws of Thought* (1854) [3] appeared "on the very day" as his own and soon overshadowed it [27]. Although Boole's work is of greater depth than De Morgan's, his initial publications on the subject were inspired, at least in part, by De Morgan's work, as he notes in the preface.

Boole introduced himself to De Morgan by letter in 1842, when Boole was not yet widely known. Excessively modest and self-conscious of the fact that he had no degree, Boole said that he had no "claim to [the] notice" of those mathematicians who did [16]. De Morgan encouraged him to submit his work to the Royal Society, assuring him that he need not hesitate a moment to publish. This award-winning paper immediately brought him to the notice of the mathematical community.

As Boole's reputation grew, he turned to De Morgan on numerous occasions to proofread his papers and arrange their publication. (Some 90 letters from the Boole–De Morgan correspondence have survived [28].) De Morgan always supplied positive reinforcement, even when busy with his own work or when he didn't understand Boole's train of thought: "To say how far I agree with you would be difficult at this time, as it is my busiest time. . . . But I must urge on you to continue and publish." On another occasion, De Morgan writes, "I hope you will expand your views of probabilities—which I am not sure I understand" [28].

De Morgan secured a pass to the library of the British Museum for Boole and was instrumental in securing a position for him at Queen's College, Cork, no small accomplishment given that the last line of Boole's application read, "I am not a member of any University and have never studied at a college." De Morgan writes in his recommendation, "I can speak confidently to the fact of his being not only well-versed in the highest branches of Mathematics, but possessed of original power" [16].

Though he enjoyed his position, Boole was not happy in Ireland, and troubles within the college worried him. With no wife or family nearby, Boole turned to De Morgan for advice and support [28]:

> Now this is what I would not say to any one in whose good feeling and discretion I could not place entire confidence. What I ask of you is not to mention these circumstances but to inform me at any future period of what you suppose might suit me in England. No one else knows of my present views and feelings.

By 1855, Boole's mood had improved and he reported to De Morgan: "My objections to Ireland are however growing less and less and I have really very little to complain of besides the smallness of the remuneration which I receive" [28]. The reason for his cheerfulness was apparent when he revealed that he had married four months earlier. At the request of a friend, Boole had agreed in 1850 to tutor Mary Everest. Boole had initially warned Mary that he was too old ever to think of marrying, but her father's death in 1855 left her ill and destitute. At the age of 40, Boole proposed to the 23-year old, cautioning that he had reservations about "imprisoning a young girl's life" [16]. By all accounts, their marriage was an extremely happy one: Mary referred to their nine years together as "a sunny dream" [6].

Boole's correspondence with De Morgan clearly indicates the depth of respect and trust he had for him. As mathematician Robert Graves, a contemporary of De Morgan, puts it, "the sterling truthfulness" of De Morgan's nature nurtured confidence in his friends, and they confided in him their most private thoughts and feelings [12]. When Boole died unexpectedly at the age of 49, De Morgan worked hard to secure a government pension for his wife and five small children. In a letter to Hamilton dated December 13, 1864, De Morgan writes, "There will be no need to tell you that you must be aiding and assisting in getting a pension for Boole's wife and daughters. An application will be made and must be well-backed" [13].

Significantly, De Morgan used distinctly different tones in his letters to Boole and to Hamilton. With Hamilton, De Morgan often assumed a bossy, even reprimanding voice. Sarcastic remarks were common: "Ink must be cheap in Ireland if you can afford to waste it on such a supposition as that" [13]. With the shy and insecure Boole, De Morgan was much gentler. His perceptiveness is remarkable, given that he was not well acquainted with either of them in person.

Gompertz

Benjamin Gompertz (1779–1865) was the son of a distinguished diamond merchant. Like Boole, he was forced to teach himself mathematics for, as a Jew, he was barred from a college education. He studied Newton and Maclaurin, stealing out into the garden and pursuing his investigations by moonlight after his parents removed his candles fearing he would stay up late studying. His first significant mathematics paper described an application of a method of differences to series. It was submitted to the Royal Society in 1806, but the two papers that established his reputation as a brilliant mathematician and won him an invitation to join the Royal Society appeared in 1817 and 1818. When the Royal Society rejected the first for being too profound to be understood, Gompertz published them at his own expense [1].

Gompertz was, as De Morgan wrote in an obituary [9],

> in a certain sense, 'the last of the learned Newtonians.' He was the last who adhered to the old language of fluxions, which [as of 1865] has been obsolete in the English mathematical world for nearly half a century.

It is adherence to fluxional notation that apparently prevented wide recognition of his work at the time. Today Gompertz is regarded as a pioneer in actuarial science.

Gompertz's wife came from a wealthy Jewish family with strong links to the stock exchange. Gompertz's brother-in-law set up the Alliance Assurance Company in 1824. Gompertz served as actuary from then until his retirement from active life. In 1825, he introduced the law of mortality, which became a fundamental tool of the life insurance industry.

De Morgan and Gompertz

Though his primary job was at University College, De Morgan frequently accepted work as an actuarial consultant. In 1838 he published *An Essay on Probabilities, and on Their Applications to Life Contingencies and Insurance Offices*, which applied probabilistic methods to insurance problems [8]. It was widely used for more than a generation.

A close personal friend and correspondent of Gompertz, De Morgan supported Gompertz when his law of mortality did not immediately capture the attention of those in the field. De Morgan wrote that it was "not by any means so well known as it ought to be, even by actuaries" [10]. Worse, it was attacked, in 1832, by T. R. Edmonds, who claimed that he had discovered the same law "independent of the imperfect one of Mr. Gompertz" [15]. De Morgan defended his friend, writing, "*All* of the points of Mr. Edmonds's alleged discovery had been published by Mr. Gompertz," and whereas Edmonds claims "that the discovery of Mr. Gompertz is imperfect—meaning, of course, as compared with that claimed by Mr. Edmonds . . . there is no difference between the two" [11]. A 15-page response from Edmonds followed, but De Morgan's defense of Gompertz was convincing.

Ramchundra

Born about 50 miles from Delhi, Yesudas Ramchundra (1821–1880) was the son of a revenue collector for the East India Company. De Morgan reports that he attended a school where "no particular attention was paid to mathematics" and "studied [mathematics] at home with such books as he could procure" [22]. One year after his father died, Ramchundra was married at the age of 11 to the daughter of a wealthy man of Delhi. Although her dowry eased the burden of supporting his mother and five siblings somewhat, his wife was deaf and mute, as Ramchundra only learned after the wedding. He was forced to drop out of school for three years to care for his new wife and allow his siblings to get an education. During this period, he worked as a journalist. Returning to Delhi College on scholarship and subsequently completing his studies in 1844, Ramchundra was hired by his alma mater to teach science and mathematics.

Ramchundra was born during a period in which the cultural and literary life of Delhi was vibrant. He was responsible for the Delhi

College publications during the 1840s and '50s, making Western developments in science and technology available to the literate public of north India and advocating openness to knowledge independent of its origin. Praised by scholars for a straightforward and conversational style of writing, he wrote on many topics apart from mathematics, including water mills, banyan trees, steamboats, railways, balloons, mirages, deception, irrigation, Confucious, the circulation of blood, and the education of girls [17].

During the Indian revolt of 1857, the principal of Delhi College was killed, and the college collapsed. In a letter to De Morgan, Ramchundra detailed his own narrow escape when mutineers raided his village [22]:

> [A] very prudent Brahmin zemindar advised me and my servant to fly to the jungles before the mutineers could arrive. We did so; but before we could run three quarters of a mile, we heard a great noise in the village, bullets were whistling about us, and horsemen appeared to be in our pursuit, for the noise of galloping was distinctly heard. I then rushed into a thorny little bush, not minding the thorns that went into my flesh . . . the mutineers, after plundering and giving a good beating to the zemindars, &c. with whom I lived in the village, did not penetrate into the jungle, but went their way towards Delhi.

After a brief appointment as headmaster of what is today the Indian Institute of Technology—Roorkee, he was appointed in 1858 as headmaster of a newly organized school in Delhi. He retired at the age of 45 because of poor health but was later appointed director of education in 1870 and honored for his contribution to the development of education in the state of Patiala.

De Morgan and Ramchundra

In 1850, De Morgan received a copy of a work by a 29-year-old, self-taught Indian mathematician. "[M]y own birth and descent having always given me a lively interest in all that relates to India, I took up the work of Ramchundra with a mingled feeling of satisfaction and curiosity: a few minutes of perusal added much to both" [22]. Like Gompertz, Ramchundra had published his treatise in Calcutta at his own expense. Endorsing the work as being worthy of encouragement, De Morgan recommended that the treatise be reprinted in London for circulation in Europe and India and offered his services as editor. *A Treatise on Problems of Maxima and Minima* appeared nine years later with a 23-page preface by De Morgan [22]. Upon De Morgan's recommendation, the Indian government authorized a reward of 2,000 rupees to a very grateful Ramchundra.

De Morgan had hoped that the reprint would bring Ramchundra to the notice of scientific men in Europe and promote a revival of interest in India. He pointed out that mathematics is one of the sciences for which Europe is indebted to India. Unfortunately, things didn't turn out as De Morgan had planned and Ramchundra's name is not widely known today, even within mathematics. Niven [20], for example, contains no mention of Ramchundra, although Ramchundra's *A Treatise on Problems of Maxima and Minima, Solved by Algebra* (1859) addresses the same ideas and even some of the same problems [22]. Rice [25] characterizes De Morgan's labor on Ramchundra's behalf as "*maximum effort, minimum effect.*" On the other hand, since the appearance of Musès' paper [18], a biography of Ramchundra has been added to various online encyclopedias (e.g., Wikepedia and Answers.com), and a 2007 publication by Nahin [19] has a section titled "Apollonius Pursuit and Ramchundra's Intercept Problem." Perhaps seeds sown by De Morgan more than a century ago finally have taken root.

Acknowledgments

The author would like to thank the editor and the reviewers for their excellent suggestions, and M. Meo for bringing De Morgan's relationship with Ramchundra to her attention. She would also like to acknowledge support received from the UCO Office of Research and Grants while conducting research for this manuscript.

References

1. M. N. Adler, Memoirs of the late Benjamin Gompertz, *Journal of the Institute of Actuaries* **13** (1866) 1–20.
2. G. Boole, Exposition of a general theory of linear transformations, *Cambridge Journal* **3** (1841) 1–20.
3. ———, *An Investigation of the Laws of Thought on Which Are Founded the Mathematical Theories of Logic and Probabilities*, Macmillan, London, 1854.

4. ⸺, *The Mathematical Analysis of Logic, Being an Essay Towards a Calculus of Deductive Reasoning*, Macmillan, Barclay, & Macmillan, Cambridge, 1847.

5. ⸺, On a general method in analysis, *Philosophical Transactions of the Royal Society of London* **134** (II) (1844) 225–282.

6. M. E. Boole, Home-side of a scientific mind, Mary Everest Boole: *Collected Works* vol. 1, C. W. Daniel Company, London, 1931.

7. R. Corrie, *Penny Magazine Online*, Electronic Historical Publications; available at http://www.history.rochester.edu/pennymag/, accessed August 23, 2009.

8. A. De Morgan, *An Essay on Probabilities, and Their Application to Life Contingencies and Insurance Offices*, Longman, Orme, Brown, Green & Longmans, London, 1838.

9. ⸺, Benjamin Gompertz, *The Athenaeum*, 22 July 1865, 117.

10. ⸺, On a property of Mr. Gompertz's law of mortality, *Journal of the Institute of Actuaries* **8** (1858–1860) 181–184.

11. ⸺, On an unfair suppression of due acknowledgment to the writings of Mr. Benjamin Gompertz, *Journal of the Institute of Actuaries* **9** (1860–1861) 86–89.

12. R. P. Graves, *Life of Sir William Rowan Hamilton*, 1882–1889, vol. 2, Arno P., New York, 1975.

13. ⸺, *Life of Sir William Rowan Hamilton* 1882–1889, vol 3, Arno P., New York, 1975.

14. T. L. Hankins, *Sir William Rowan Hamilton*, John Hopkins UP, Baltimore, 1980.

15. P. F. Hooker, Benjamin Gompertz, *Journal of the Institute of Actuaries* **91** (1965) 203–212.

16. D. MacHale, *George Boole: His Life and Work*, Boole P., Dublin, Ireland, 1985.

17. G. Minault, Master Ramchandra of Delhi College: teacher, journalist, and cultural intermediary, *The Annual of Urdu Studies* **18** (2003) 95–104; available at http://www.urdustudies.com/pdf/18/10MinaultRamchandra.pdf, accessed August 23, 2009.

18. C. Musès, De Morgan's Ramanujan: an incident in recovering our endangered cultural memory of mathematics, Math. *Intelligencer* **20**(3) (1998) 47–51. doi:10.1007/BF03024806.

19. P. J. Nahin, *In Chases and Escapes: The Mathematics of Pursuit and Evasion*, Princeton UP, Princeton, NJ, 2007.

20. I. Niven, *Maxima and Minima Without Calculus*, Mathematical Association of America, Washington, DC, 1981.

21. S. O'Donnell, *William Rowan Hamilton: Portrait of a Prodigy*, Boole P., Dublin, Ireland, 1983.

22. Ramchundra, *A Treatise on Problems of Maxima and Minima, Solved By Algebra*, Wm. H. Allen, London, 1859.

23. A. Rice, Augustus De Morgan: Historian of science, *Hist. Sci.* **34** (1996) 201–240.

24. ⸺, Inspiration or desperation? Augustus De Morgan's appointment to the chair of mathematics at London University in 1828, *British J. Hist. Sci.* **30** (1997) 257–274. doi:10.1017/S0007087497003075.

25. ⸺, Maximum effort, minimum effect: De Morgan and his Indian protégé, "History of Mathematics: Mathematics in the Americas and the Far East, 1800–1940," conference presentation, October 1998; abstract available at http://www.mfo.de/programme/schedule/1998/43/Report_41_98.pdf, accessed August 23, 2009.

26. ⸺, What makes a great mathematics teacher? The case of Augustus De Morgan, *American Mathematical Monthly* **106** (1999) 534–552. doi:10.2307/2589465.

27. ⸺, Everybody makes errors: The intersection of De Morgan's logic and probability, 1837–1847, *Hist. Philos. Logic* **24** (2003) 289–305. doi:10.1080/01445340310001599579.

28. G. C. Smith, *The Boole–De Morgan Correspondence 1842–1864*, Oxford, New York, 1982.

Routing Problems:
A Historical Perspective

GIUSEPPE BRUNO, ANDREA GENOVESE,
AND GENNARO IMPROTA

From the Königsberg Bridges
to the Chinese Postman Problem

In 1741, Leonhard Euler published (in the *Commentarii* of the Saint Petersburg Academy) a paper presenting some results related to the so-called Seven Bridges of Königsberg Problem. The Pregel river (*Pregolja* in Russian), coming from the east, crosses Lithuania and enters a Russian enclave (once named Eastern Prussia, between Lithuania and Poland) whose main city is Kaliningrad (the ancient Königsberg). The two branches of the river (*Novaya Pregolja* and *Staraya Pregolja*) cross Königsberg, forming an island in the heart of the city before merging and leading to Vistula Lagoon and then to the Baltic Sea.[1] Königsberg city center was composed of four main areas (west bank, east bank, central island, and a small quarter surrounded by the two branches of the Pregel River) connected together by seven bridges, some of them still existing (Mallion 2008).

Taking up a popular anecdote of the time, Euler formulated the following problem (authors' translation):

> In the Prussian city of Königsberg there is an island called Der Kneiphof; around this island, the two branches of the [Pregel] river flow. There are seven bridges crossing the two branches. About these bridges, this question was asked: is it possible to build a route to pass through each of the bridges once but not more than once? I was told that some denied and others doubted that this could be done, but nobody took it for sure. From this, I

have developed this general problem: whatever the configuration and distribution of river branches and whatever the number of bridges, can you find out if you can pass through each bridge once and only once?

Euler proposed a mathematical formulation of the problem, which is usually regarded as the birth of the graph theory, and showed that such a tour across Königsberg bridges was not feasible. Turning to a more general situation, he introduced some necessary and sufficient conditions for the existence of what is now called an Eulerian circuit.[2] Almost incidentally, he started to talk about routing problems. He did not create an algorithm for finding such a tour; his only objective was to determine the existence of a tour, without necessarily identifying it.

In 1962 (221 years later), in Mao Tse Tung's China, Mei-Ko Kwan, a mathematician who had worked as a postman during his youth, described his idea of extending Euler's problem as follows:

Suppose there is a mailman who needs to deliver mail to a certain neighbourhood. This mailman is lazy, so he wants to find the shortest route through the neighbourhood, that meets the following criteria: it has to be a closed circuit (it ends at the same point it starts); the route has to go through every street at least once.

Alan Goldman, a researcher working at the National Institute of Standards and Technology in the United States, inspired by the nature of the problem and the nationality of the scholar who proposed it first, coined the term Chinese postman problem (*CPP*), still accepted and widespread today.[3] Goldman suggested his idea to Jack Edmonds (among the founding fathers of combinatorial optimization as a field of study), who started to refer to the problem this way.

We must point out that Mei-Ko Kwan's research question is broader than Euler's, as it looks for the construction of a circuit also in cases where the graph representing the problem is not Eulerian (that is, it has more than two odd vertices). In such cases, the CPP can be solved by modifying the graph to an Eulerian version through appropriate algorithms. Euler's considerations are therefore crucial for developing algorithms for solving the CPP as well.

The seven bridges of Königsberg problem and its extension to the Chinese postman problem constitute the starting point for a class of arc

Figure 1. Chinese mail carrier, from *History of the World's Postal Service*, 1886–1888.

routing problems that deal with pickup and delivery problems across arcs constituting paths and routes.

The Icosian Game and the Traveling Salesman Problem

In 1832, in Germany, a manual entitled *Der Handlungsreisende* was devoted to the traveling salesman. In it are the following sentences (translated in Schrijver 2003):

> Business brings the traveling salesman now here, then there, and no travel routes can be properly indicated that are suitable for all cases occurring; but sometimes, by an appropriate choice and arrangement of the tour, so much time can be gained, that we don't think we may avoid giving some rules also on this. Everybody may use that much of it as he takes it for useful for his goal; so much of it however we think we may assure, that it will not be well feasible to arrange the tours through Germany with more economy in view of the distances and, which the traveler mainly has to consider, of the trip back and forth. The main point always

consists of visiting as many places as possible, without having to touch the same place twice.

This booklet, without any mathematical calculation, suggested five possible tours among 45 German cities, based only on empirical considerations. In a paper published in 1983, Heiner Müller-Merbach recognizes in these sentences a first, rough, description of the traveling salesman problem (TSP).

In England in the mid-nineteenth century, two mathematicians developed relevant theoretical contributions to the problem. In a paper that appeared in 1856, Thomas Penyngton Kirkman, rector of Southworth in Lancashire, proposed the first graphic formulation of the TSP as follows:

> Given the graph of a polyhedron, is it possible to obtain a final circuit stopping by each node once and only once?

In 1857, the Irish mathematician William Rowan Hamilton launched a somewhat eccentric initiative: He circulated a board game called the Icosian Game, based on a regular dodecahedron, with the name of a city associated with each edge. The commercial rights of the Icosian Game were acquired (for £25) by a games wholesaler (named Jaques) who started to sell it with the captivating name of "The Travellers Dodecahedron or A Voyage Around the World." The game was manufactured in different versions (endowed with different optional features for different markets) and proved to be quite a successful initiative.

Figure 2 shows a planar version of the Icosian Game, based on a schematic representation of the edges and vertices of a dodecahedron. Each vertex (corresponding to a city to be visited) is marked by a small cavity, in which pins can be plugged to keep track of the tour; a silk thread is provided to keep track of the route.

The basic game consisted in finding a tour that starts from a city (a vertex) and visits all the other cities once and only once before returning to the starting point. However, many variants of the game were proposed.[4]

In Hamilton's game, as in Euler's problem, the main objective is the construction of a tour. Their solutions are therefore associated with a routing problem. The two problems have significant conceptual differences, however: In the Königsberg bridges problem, Euler was looking for a tour crossing all the arcs of a graph once and only once; Hamilton,

Figure 2. Hamilton's Icosian game.

on the other hand, asked for a tour stopping by all the nodes of the graph once and only once (a Hamiltonian path).[5] Thus, the two problems complement each other.

The first mathematical formulation of the TSP was delivered by the Austrian mathematician Karl Menger who around 1930 worked at Vienna and Harvard. Menger originally named the problem the messenger problem,[6] and set out the difficulties as follows (Menger 1932, translated in Bock 1963). At this time, computational complexity theory had not yet been developed:

> We designate as the Messenger Problem (since this problem is encountered by every postal messenger, as well as by many travelers) the task of finding, for a finite number of points whose pairwise distances are known, the shortest path connecting the points. This problem is naturally always solvable by making a finite number of trials. Rules are not known which would reduce the number of trials below the number of permutations of the given points. The rule, that one should first go from the starting point to the point nearest this, etc., does not in general result in the shortest path.

The Traveling Salesman Problem after World War II

In the 1950s and 1960s, thanks to the first computers and to theoretical development in the field of combinatorial optimization, the interest of scholars and researchers in the TSP increased, both in Europe and the United States. A significant contribution came from the team formed by George B. Dantzig, Delbert Ray Fulkerson, and Selmer M. Johnson, based at the Research and Development (RAND) Corporation in Santa Monica. In 1954, they proposed an integer linear programming formulation. Moreover, they developed the cutting-plane method, which enables the finding of an optimal solution (namely, the shortest Hamiltonian tour) for a TSP involving the 49 U.S. state capitals.

It is worth highlighting that the simplicity of the definition and formulation of the TSP can be misleading. Indeed, it is easy to compute the total number of possible tours among n cities: Once a starting city has been fixed, there are clearly $(n-1)!$ possible tours (if the distances are symmetric, this number reduces to $(n-1)!/2$). The presence of the factorial gives rise to a rapid growth in the number of feasible solutions as the number of nodes increases. Menger's problem (to reduce the number of trials to fewer than this) has not yet been resolved; however, it has been possible to solve the TSP for an increasing number of nodes (or "instances") by using heuristic or program-based approaches and increased computing power.

In 1975, Camerini, Fratta, and Maffioli solved the TSP for 67 cities, a feat that was unbeaten for two years. In 1987 another Italian team, Rinaldi and Padberg, improved on this twice, to 532 cities, and then to 2,392. These improvements were made using branch-and-cut algorithms on parallel machines (Padberg and Rinaldi 1991). In 2001, Applegate, Bixby, Chvátal, and Cook solved the TSP for 15,112 German cities, using a network of parallel workstations at Rice University and Princeton (Ciriani 2001). Three years later, Applegate *et al.* (2004) went up to 24,978 nodes, representing Swedish villages and cities, by building a 72,500-km (44,740-mi) tour and showed that it is not possible to find a shortest tour. It is commonly thought that the development of new algorithms and faster computers will enable solutions up to 100,000 nodes (Applegate *et al.* 2006). Figure 3 depicts the evolution of solutions of the TSP since 1954.

Some Extensions: The Vehicle Routing Problem

Besides the basic TSP, other routing problems have been formulated with the objective of reproducing more realistic logistic problems; often,

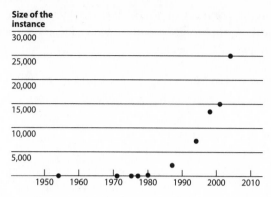

Figure 3. Evolution of the solution of the TSP.

these problems are variants of the original TSP, obtained by adding one or more constraints. In the version known as TSP Windows (TSPW), for example, the salesman has to visit each node within a certain time frame. Another extension to the TSP is the so-called multiple-TSP, commonly referred to as m-TSP. In this version, m traveling salesman have to visit n cities, and every city must be visited once and only once by each salesman. All the salesmen depart from a depot (which need not be the same for each of them) and must return to it at the end.

If salesmen are replaced by vehicles (each with a given capacity), and a demand for goods is associated with each city, the m-TSP is transformed into the vehicle routing problem (VRP). The optimization criterion for the VRP is usually the minimization of the total distance covered; sometimes, however, it is the number of vehicles m that must be minimized. It is easy to understand that the VRP provides a strong methodological basis for supply chain management (SCM) applications.

The seminal contribution to the VRP can be found in a brief paper by Dantzig and Ramser (1959). The authors introduced the problem in the following way:

> The paper is concerned with the optimum routing of a fleet of gasoline delivery trucks between a bulk terminal and a large number of service stations supplied by the terminal. . . . It is desired to find a way to assign stations to trucks in such a manner that station demands are satisfied and total mileage covered by the fleet is a minimum. A procedure based on a linear programming

formulation is given for obtaining a near optimal solution. The calculations may be readily performed by hand or by an automatic digital computing machine.

The first algorithm for solving the VRP was proposed by Clarke and Wright (1964), the former a researcher at the College of Science and Technology of the University of Manchester, the latter a technician of the Manchester Cooperative Wholesale Society. In a joint research program launched by the two institutions, they proposed an effective heuristic procedure capable of significantly improving the performances of the Dantzig and Ramser approach. The abstract of their paper ended with this sentence:

> This paper . . . develops an iterative procedure that enables the rapid selection of an optimum or near-optimum route. It has been programmed for a digital computer but is also suitable for hand computation.

The first paper referring to this problem as the vehicle routing problem was published by Bruce L Golden, Thomas L Magnanti, and Hien Quang Nguyen (1972). However, some variants had already been proposed during the early 1970s. Liebman (1970) had formulated a pickup and delivery problem for the urban solid waste management cycle. Levin (1971) had introduced the fleet routing problem for an urban public transport system; Wilson and Sussman (1971) had used the dial-a-ride problem (DARP, a VRP variant) for on-demand public transport systems. The VRP variants can be considered part of a more general problem, the general pickup and delivery problem (GPDP).

Research on VRP and its variants became increasingly popular during the 1990s: This fact is testified to by the growing number of papers published in international refereed journals. Between 2000 and 2006 (Figure 4), 447 contributions on VRP and its variants were published.

Some Applications

The ideas of Euler and Hamilton constitute the seminal elements of a research area that is still in evolution. In particular, the ideas underlying routing problems have been applied to tackle and solve a variety of problems even in fields apparently unrelated to the field of logistics and

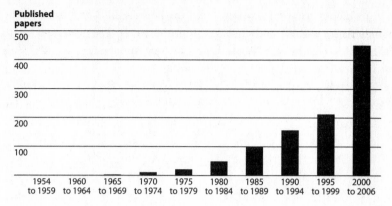

Figure 4. Growth of the literature on VRP (Eksioglu et al. 2009).

supply chain management. Indeed, both the CPP (and its several extensions allowing, for example, for capacitated arcs and time-dependent costs) and the TSP (and its main variant, the VRP) have inspired models and methods for dealing with a wide range of logistic problems (e.g., transportation planning, waste cycle management systems, road cleaning, and broadcasting line inspections).

One of the first applications of the TSP can be traced back to 1940. The Indian Prasanta Chandra Mahalanobis was in charge of conducting a census of jute crops in the Ganges Delta on behalf of the Department of Agriculture of West Bengal. He noticed that one of the most relevant costs of the survey was the expense of transporting men and machinery to the different places to be inspected. Mahalanobis planned the activities using a TSP-based model, thus achieving a marked economic benefit (Mahalanobis 1940). Some years later, in 1948, the U.S. mathematician Merrill M. Flood (one of the main contributors to game theory) developed an adaptation of the TSP model for solving an urban transportation planning problem. Flood (1956) (at the time working at the RAND Corporation) contributed to the diffusion of the TSP and promoted research in this field of study.

Though transportation of people and goods has been the main application of the TSP, the flexibility of the model has also allowed adaptation to other disciplines. A classical example is provided by the electronic devices industry and, in particular, by the production process for integrated circuit components. The TSP is widely used for programming the machines used for drilling holes in the boards: The holes are the nodes to be visited, and costs are represented by the time needed to move the mechanical drill arm from one hole to the next. By 2005, this problem had been solved for 33,810 points.

Researchers from the National Institutes of Health in Bethesda, Maryland, and from the Department of Computational and Applied Mathematics at Rice University used the TSP to produce maps of hybrid radiations during studies on the human genome (Agarwala *et al.* 2000). The method allowed the bringing together of local maps into a single global map. In this adaptation of the TSP, local maps represent the nodes; the cost is calculated from the likelihood that a partial map *i* immediately follows another map *j* in composing the map of the genome. A few years later, a similar algorithm was used to calculate DNA sequences (Avner *et al.* 2001).

In 2001, a research team at Hernandez Engineering Inc. (in Houston) and at Brigham Young University (in Provo, Utah) used the Lin–Kernighan heuristic algorithm to determine the sequence of stellar bodies to be visited by two spacecraft in the StarLight space interferometer mission (for the acquisition of images), in order to minimize the fuel consumption of the two space probes (Bailey *et al.* 2000). The analysis of crystal structure (Bland and Shallcross 1987), the management of materials in warehouses (Ratliff and Rosenthal 1981), the classification of databases (Lenstra and Rinnooy Kan 1975), the sequence of operations on a machine (Gilmore and Gomory 1964), the revision of turbine engines (Plante *et al.* 1987), and the sequencing of takeoffs and landings of aircraft (Agarwal *et al.* 2004) are some other examples of numerous applications of the TSP.

Notes

1. Some details about Königsberg city life during the eighteenth century can be found in the novel *Critica della Ragion Criminale* by Michael Gregorio (alias Michael Jacob and Daniela De Gregorio 2006). One of the main characters of the novel is the philosopher Immanuel Kant.

2. If the graph that represents the problem is connected and all the vertices have even degree, an Eulerian circuit exists. Euler also showed that a path exists if the graph has only two vertices with odd degree, in which case the path must depart from one of these two vertices.

3. This is also sometimes known as the route inspection problem (RIP). Goldman affirms, in a personal note dated 14 December 2003, that he was influenced in selecting the name of the problem by the title of a noir novel by Ellery Quen, *The Chinese orange mystery* (1934).

4. According to the original rules, up to 15 different games could be implemented. Hamilton's idea was that two people could compete against each other, the first by selecting the version of the game and setting up the rules, the second by solving it. Hamilton (1856, 1858) also developed a discipline that he named Icosian calculus, which he used to investigate closed paths on dodecahedrons.

5. When a Hamiltonian path has an arc that connects the last vertex to the first one, it is called a Hamiltonian circuit.

6. Morton and Land (1955) state that the TSP was originally named the laundry van problem.

Bibliography

Agarwal, A., Lim, M. H., Chew, C. Y., Poo, T. K., Er, M. J., and Leong, Y. K. "Solution to the fixed airbase problem for autonomous URAV site visitation sequencing," *Genetic and evolutionary computation*, Springer, 2004.

Agarwala, R., Applegate, D. L., Maglott, D., Schuler, G. D., and Schaffer, A. A., "A fast and scalable radiation hybrid map construction and integration strategy," *Genome Research*, 10 (2000), 350–364.

Applegate, D., Bixby, R., Chvátal, V., Cook, W., *The German cities TSP solution*. http://www.tsp.gatech.edu/d15sol/dhistory.html, 2001 (accessed Nov. 24, 2010).

Applegate, D., Bixby, R., Chvátal, V., and Cook, W., *The traveling salesman problem: A computational study*, Princeton University Press, 2006.

Applegate, D., Bixby, R., Chvátal, V., Cook, W., and Helsghaun, K., *The Sweden cities TSP solution*. http://www.tsp.gatech.edu//sweden/cities/cities.htm, 2004 (accessed Nov. 24, 2010).

Avner, P., et al., "A radiation hybrid transcript map of the mouse genome," *Nature Genetics*, 29 (2001), 194–200.

Bailey, A. C., McLain, T. W., and Beard, R. W., "Fuel saving strategies for separated spacecraft interferometry," *Proceedings of the AIAA Guidance, Navigation, and Control Conference*, 2000.

Bland, R. E., and Shallcross, D. F., *Large traveling salesman problem arising from experiments in X-ray crystallography: A preliminary report on computation*, Technical Report No. 730, School of OR/IE, Cornell University, 1987.

Bock, F., "Mathematical programming solution of traveling salesman examples," *Recent advances in mathematical programming*, R. L. Graves and P. Wolfe (eds.), McGraw-Hill, 1963.

Bruno, G., Genovese, A., Improta, G., "Storia dei problemi di routing, da Eulero ai nostri giorni," *Proceedings of the 3rd Italian Conference on Engineering History* (in Italian), 2010.

Camerini, P. M., Fratta, L., and Maffioli, F., "On improving relaxation methods by modified gradient techniques," *Mathematical programming study*, 3 (1975), 26–34.

Ciriani, T. A., "4>15112," *AIROnews VI*, 4 (2001), 9–10.

Clarke, G., and Wright, J. W., "Scheduling of vehicles from a depot to a number of delivery points," *Operations Research*, 12 (1964), 568–581.

Dantzig, G., and Ramser, J. H., "The truck dispatching problem," *Management Science*, 6 (1959), 80–91.

Dantzig, G. B., Fulkerson, D. R., and Johnson, S. M., "Solution of a large scale traveling salesman problem," *Journal of the Operations Research Society of America*, 2 (1954), 393–410.

[ein alter Commis-Voyageur'], *Der Handlungsreisende—wie er sein soll und was er zu thun hat, um Aufträge zu erhalten und eines glücklichen Erfolgs in seinen Geschäften gewiss zu sein—Von einem alten Commis-Voyageur*, B Fr Voigt, Ilmenau, 1832; reprinted Verlag Bernd Schramm, Kiel, 1981.

Eksioglu, B., Vural, V., and Reisman, A., "The vehicle routing problem: A taxonomic review," *Computers & Industrial Engineering*, 57 (2009), 1472–1483.

Euler, L., "Solutio problematis ad geometriam situs pertinentis," *Commentarii Academiae Scientiarum Imperialis Petropolitanae*, 8 (1741), 128–140, the original text is available at http://www.math.dartmouth.edu/~euler/ (accessed Nov. 24, 2010).

Flood, M. M., "The traveling-salesman problem," *Operations Research*, 4 (1956), 61–75.

Gilmore, P. C., and Gomory, R. E., "Sequencing a one state-variable machine: A solvable case of the traveling salesman problem," *Operational Research*, 12 (1964), 655–679.

Golden, B. L., Magnanti, T. L., and Nguyen, H. Q., "Implementing vehicle routing algorithms," *Networks*, 7 (1972), 113–148.

Gregorio, M., *Critica della ragion criminale*, Einaudi, Turin, Italy, 2006.

Hamilton, W. R., "Memorandum respecting a new system of roots of unity (the Icosian calculus)," *Philosophical Magazine*, 12 (1856), 446.

Hamilton, W. R., "Memorandum respecting a new system of roots of unity (the Icosian calculus)," *Proceedings of the Royal Irish Academy*, 6 (1858), 415–416.

Kirkman, T. P., "On the representation of polyhedra," *Philosophical Transactions of the Royal Society*, 146 (1856), 413–418.

Lenstra, J. K., and Rinnooy Kan, A.H.G., "Some simple applications of the travelling salesman problem," *Operational Research Quarterly*, 26 (1975), 717–733.

Levin, A., "Scheduling and fleet routing models for transportation systems," *Transportation Science*, 5 (1971), 232–256.

Liebman, J. C., "Mathematical models for solid waste collection and disposal," *Bulletin of the Operations Research Society*, 18 (1970).

Mahalanobis, P. C., "A sample survey of the acreage under jute in Bengal," *Sankhyā*, 4 (1940), 511–530.

Mallion, R., "A contemporary Eulerian walk over the bridges of Kaliningrad," *BSHM Bulletin*, 23 (2008), 24–36.

Mei-Ko Kwan, "Graphic programming using odd or even points," *Chinese Mathematics*, 1 (1962), 273–277.

Menger, K., "Eine neue Definition der Bogenlange," *Ergebnisse eines Mathematischen Kolloquiums*, 2 (1932), 11–12.

Morton, G., and Land, A. H., "A contribution to the travelling-salesman problem," *Journal of the Royal Statistical Society*, 17 (1955), 185–194.

Müller-Merbach, M., "Zweimal travelling salesman," *DGOR-Bulletin*, 25 (1983), 12–13.

Padberg, M., and Rinaldi, G., "A branch-and-cut algorithm for the resolution of large-scale symmetric traveling salesman problems," *SIAM Review*, 33 (1991), 60–100.

Plante, R. D., Lowe, T. J., and Chandrasekaran, R., "The product matrix traveling salesman problem: An application and solution heuristic," *Operations Research*, 35 (1987), 772–783.

Queen, E. *The Chinese orange mystery*, Stokes, New York, 1934.

Ratliff, H. D., and Rosenthal, A. S., *Order picking in a rectangular warehouse: A solvable case for the traveling salesman problem*, PDRC Report Series No. 81-10, Georgia Institute of Technology, Atlanta, 1981.

Schrijver, A., *Combinatorial optimization: Polyhedra and efficiency*, Vol. 1, Springer-Verlag, Berlin, Heidelberg, 2003.

Wilson, N.H.M., and Sussman, J. M., *Scheduling algorithms for dial-a-ride systems*, Urban Systems Laboratory Report USL TR-70-13, MIT, Cambridge, MA, 1971.

The Cycloid and Jean Bernoulli

GERALD L. ALEXANDERSON

Johann (Jean) Bernoulli (1667–1748), the younger brother of Jacob (Jacques) Bernoulli, was a member of a large family of respected spice traders and scholars in Basel. Originally Flemish, they had fled to Switzerland to avoid religious persecution at the hands of the Spanish, who then occupied the Low Countries.

The 1742 *Opera Omnia* of Johann is a four-volume set that includes not only his mathematics but also work on fermentation and on the design of naval vessels. Handsomely produced by the Swiss firm of Bousquet, the first volume opens with a frontispiece portrait of Bernoulli (Figure 1) followed by a colorful title page (Figure 2). On that page there is a curious engraving of a dog with its front feet on the trunk of a tree—a palm, perhaps?—in a mountainous landscape, with the dog looking at a picture of a geometric figure on a large scroll attached to the tree. As it turns out, the figure is a cycloid. Similarly, on the facing page, there is the extravagantly elaborate oval engraving of the bust of Bernoulli. He is holding a rolled piece of paper showing the same figure, which appears as Figure 1 of Table XV, facing page 336 in volume 1. The illustration is used in finding the center of gravity of a sector of a cycloid and appeared in an article in the *Acta Eruditorum* of June of 1699, page 316. The figure also appears on page 609 of the award-winning article by Apostol and Mnatsakanian on cycloidal areas [1]. These illustrations suggest the large role this curve played in the work of Bernoulli. Just to add an element of extravagance, the dedication of the *Opera Omnia* is accompanied by a lavish engraving, if anything, outdoing the one of Bernoulli himself—but this time of the patron "Fridericus III Rex Borussiae" (Prussia). Something seems wrong, however: Frederick III of Prussia was not king until the late 19th century, long after the publication of the Bernoulli work. It would have made sense were it

Figure 1. Frontispiece of Jean Bernoulli's *Opera Omnia*, 1742.

Frederick II (the Great) who became king in 1740, two years before the publication date of the *Opera Omnia*. It's a mystery.[1] Perhaps engravings were interchangeable at that time: one picture fit all subjects. Much earlier, this would not have been unusual. The famous Nuremberg Chronicle of 1493 was generously illustrated with pictures of cities throughout Europe. But sometimes the same picture was used for more than one city! It saved money.

The publisher, Marc-Michel Bousquet, appears to have been dazzled by titles (in spite of his being Swiss). There are acknowledgments of patronage not only by the King of Prussia but also by Emperor Charles VII of the Holy Roman Empire, and Friedrich August, King of Poland and Elector of Saxony.

Calculus students recognize the cycloid as the solution to the brachistochrone problem, the curve that allows a bead rolling down a trough in the form of the curve to reach the lowest point on the curve in the least amount of time. And, surprisingly, it is also the curve on which a bead rolling down from any point on the left side of the inverted arch of a cycloid reaches the bottom at exactly the same time as a bead rolling from any point on the opposite side—that is, it is also the solution to the tautochrone problem, the curve providing the "same time descent." In the other problem—the brachistochrone—the cycloid is the curve of "least time descent." These and similar problems were of widespread interest in the late 17th and early 18th centuries.

The cycloid is, of course, the curve traced out by a point on the boundary of a circlular disk as the disk is rolled, without slipping, along a straight line. Ordinarily, it would be shown as a series of arches, though in the two contexts above, the curve has been reflected about the horizontal straight line. John Wallis thought the curve was known as early as 1451 to Nicholas de Cusa, but that claim has been questioned [6]. It was Galileo who gave it the name "cycloid" and investigated some of its properties around 1599. He was looking for curves of least time descent, though without much success. At the same time, Mersenne, Roberval, and Torricelli became interested in the curve. Pascal made some real contributions to the subject, primarily in calculating the length of the curve and various volumes as it is rotated about axes (though he often used the French word for it, *roulette*). Torricelli correctly found the area under one arch to be three times the area of

J O H A N N I S

BERNOULLI,

M. D. MATHESEOS PROFESSORIS,
Regiarum Societatum PARISIENSIS, LONDI-
NENSIS, PETROPOLITANÆ,
BEROLINENSIS, *Socii* &c.

OPERA OMNIA,

TAM ANTEA SPARSIM EDITA,
quam hactenus inedita.

TOMUS PRIMUS,
Quo continentur ea
Quæ ab ANNO 1690 ad ANNUM 1713 prodierunt.

LAUSANNÆ & GENEVÆ,

Sumptibus MARCI-MICHAELIS BOUSQUET & Sociorum.

MDCCXLII.

Cum Privilegio Sacræ Cæsareæ Majestatis , & Sereniss. Poloniæ Regis ,
Elect. Saxon.

Figure 2. Title page of Jean Bernoulli's *Opera Omnia*, 1742.

the generating circular disk. In a series of papers starting with "Problemata de Cycloide proposita mense Junii 1658" [5], along with correspondence between Huygens and A. Dettonville (a pseudonym used by Pascal—an anagram of Louis de Montalte, the name under which he published his *Lettres provinciales*), Pascal proved various results and published a challenge to others to replicate them. In 1658, Christopher Wren correctly calculated the arc length of one arch, 8*a*, where *a* is the radius of the generating circle.

The first truly impressive achievements came along later. In 1673, the Dutch physicist-mathematician, Christiaan Huygens, published his book, *Horologium Oscillatorium*, a landmark in the history of science, in which he used the tautochrone (= isochrone) property of the cycloid. Huygens designed a pendulum clock that would keep good time by having the bob of the pendulum follow the path of a cycloid so that the clock became less dependent on being placed on a horizontal surface—a result particularly applicable to clocks on a ship subject to roll when out at sea. With the importance of the spice trade at that time, having a clock on a ship that could tell time accurately was a significant motivating force in applied mathematics. An interesting picture of Huygens's clock appears on page 413 of [3], and a photograph of such a clock (identified as a "Slingerklok naar Chr. Huygens") appears on a Dutch postage stamp.

Pascal's challenge to his colleagues was not the most memorable one involving the cycloid, however. That challenge appeared in 1696 when Johann Bernoulli succeeded in showing that the cycloid is the solution of the brachistochrone problem, providing a faster descent between two points than Galileo's earlier straight line or broken line, or any other curve. Bernoulli sent a challenge (to prove that the cycloid is the solution) [2] to his older brother, Jacob, as well as to Leibniz, Newton, and l'Hôpital. Though Jacob Bernoulli was quick to respond to the challenge, his solution in finding the cycloid to be the curve of least time descent was rather clumsy and was not admired by his brother Johann, whose solution displayed some striking insights. But in the end, Jacob was the winner: His method widened the scope of the field to include isoperimetric problems and pointed in the direction of the development of the calculus of variations. (Relations between the two brothers—due to jealousy, perhaps?—were often strained.) Newton claimed that he did the brachistochrone problem in a few hours, but he is also reported to have grumbled that he did not like very much being teased by foreigners.

The years after Johann Bernoulli's important result were filled with further activity involving the cycloid. Echoing Huygens's result that the involute of a cycloid is again a cycloid but shifted along the horizontal line, it was shown that the evolute (locus of the centers of curvature) is also a congruent cycloid. The caustic of a cycloid is another cycloid (not congruent). And then there are the variants of the definition of the cycloid—the circle can roll around on the outside of the second circle (the epicycloid) or the inside (the hypocycloid). Investigations of all of these continued to interest Bernoulli, and beautiful engravings of them appear in tables LIX to LXVIII in volume 3 of his *Opera Omnia*. There are even figures indicating that he was also curious about the generalizations achieved by rolling a regular polygon around the outside of another regular polygon, generating curves that have the general appearance of an epicycloid, but are not smooth and are indeed rather lumpy! It was clearly an age of cycloids. This fascination continued for many years, and there is a reference to the cycloid even in classic American literature: Herman Melville's *Moby Dick* (1851). Galileo had remarked that part of a cycloidal arch would make a nice-looking bridge. There are claims that the repeated vaults in the ceiling of Louis Kahn's Kimball Art Museum in Fort Worth, Texas, are cycloidal. Whatever they are, they are pleasing to look at. A large demonstration piece for the brachistochrone problem played a role in the science area of the Golden Gate International Exposition in San Francisco in 1940 [6]; science museums today have similar exhibits. And a small version of the model sits on the desk in my office.

Euler's name has not played a role in this account so far, but Johann Bernoulli was well known to have been Euler's teacher. So it should come as no surprise that Euler's first publication [4], written when he was 18, opens with: "It has been observed amongst geometers that the ordinary cycloid is an isochronous or tautochronous curve. . . . I marvel greatly that no one has yet considered hypotheses for the isochrones in media with other forms of resistance" (translation by I. Bruce). Euler then proceeded to solve that problem, but some of Euler's assumptions were incorrect, and that affected the final conclusion. So even the greatest of all 18th century mathematicians could make a mistake, at least when he was very young.

Euler went on to publish a masterpiece, his *Methodus inveniendi lineas curvas* (1744), which, with the later work of Lagrange, established a whole new field of mathematics, the calculus of variations. The results on cycloids obtained before Bernoulli were for the most part done with Archimedean geometric methods, without the benefit of the calculus of Leibniz (1684) and Newton (1687). That all of the mathematicians mentioned, from Galileo up through Fermat, Descartes, Mersenne, Torricelli, and Roberval, found problems about the cycloid difficult is not surprising, but now these problems are standard exercises in calculus textbooks. It tells us something of the power of calculus.

Note

1. After publication the author heard from Olaf Teschke of Zentralblatt MATH, who cleared up the "mystery" mentioned at the top of page 211. There were many Fredericks during the time period of the article. The Frederick shown in the Bernoulli *Opera Omnia* was Frederick III, Elector of Brandenburg who later became Frederick I King of Prussia. He retained the name Frederick but the number changed when he became king, but just to complicate things further many continued to use the earlier number: Fridericus III Rex Borussiæ. I'm grateful to him for solving the "mystery."

References

[1] Apostol, Tom M., and Mnatsakanian, Mamikon A., *New insight into cycloidal areas*, Amer. Math. Monthly, 116 (2009), 598–611. MR2549377 (2010h:51019).

[2] Bernoulli, Johann, *Problema novum*, Acta Eruditorum, June 1696, p 264; *Opera Omnia*, Bousquet, Lausanne and Geneva, 1742, volume 1, p 161, Table VIII, No. XXX, Figure 4. MR0256841 (41:1791).

[3] Boyer, Carl B. (revised by Uta C. Merzbach), *A History of Mathematics, 2nd edition*, John Wiley & Sons, New York, 1989, pp 366, 374–379, 383. MR0234791 (38:3105).

[4] Euler, Leonhard, *Constructio linearum isochronarum in medio quocunque resistente*, Acta Eruditorum, 1726, pp 361–363.

[5] Pascal, Blaise, *Œuvres de Blaise Pascal*, Chez Detune, The Hague, 1779, volume 5, pp 135–225.

[6] Whitman, E. A., *Some historical notes on the cycloid*, Amer. Math. Monthly, 50 (1943), 309–315. MR0007720 (4:181p).

Was Cantor Surprised?

FERNANDO Q. GOUVÊA

Mathematicians love to tell each other stories. We tell them to our students, too, and they eventually pass them on. One of our favorites, and one that I heard as an undergraduate, is the story that Cantor was so surprised when he discovered one of his theorems that he said, "I see it, but I don't believe it!" The suggestion was that sometimes we might have a proof, and therefore *know* that something is true, but nevertheless still find it hard to believe.

That sentence can be found in Cantor's extended correspondence with Dedekind about ideas that he was just beginning to explore. This article argues that what Cantor meant to convey was not really surprise, or at least not the kind of surprise that is usually suggested. Rather, he was expressing a different, if equally familiar, emotion. To make this clear, we will look at Cantor's sentence in the context of the correspondence as a whole.

Exercises in myth-busting are often unsuccessful. As Joel Brouwer says in his poem "A Library in Alexandria,"

> . . . And so history gets written
> to prove the legend is ridiculous. But soon the legend
> replaces the history because the legend is more interesting.

Our only hope, then, lies in arguing not only that the standard story is false, but also that the real story is more interesting.

The Surprise

The result that supposedly surprised Cantor was the fact that sets of different dimension could have the same cardinality. Specifically, Cantor showed (of course, not yet using this language) that there was a bijection between the interval $I = [0, 1]$ and the n-fold product $I^n = I \times I \times \ldots \times I$.

There is no doubt, of course, that this result is "surprising," i.e., that it is counterintuitive. In fact, Cantor said so explicitly, pointing out that he had expected something different. But the story has grown in the telling, and in particular Cantor's phrase about seeing but not believing has been read as expressing what we usually mean when we see something happen and exclaim, "Unbelievable!" What we mean is not that we actually do not believe, but that we find what we know has happened to be hard to believe because it is so unusual, unexpected, surprising. In other words, the idea is that Cantor felt that the result was hard to believe even though he had a proof. His phrase has been read as suggesting that mathematical proof may engender rational certainty while still not creating intuitive certainty.

The story was then coopted to demonstrate that mathematicians often discover things that they did not expect or prove things that they did not actually want to prove. For example, here is William Byers [2, p. 179] in *How Mathematicians Think*:

> Cantor himself initially believed that a higher-dimensional figure would have a larger cardinality than a lower-dimensional one. Even after he had found the argument that demonstrated that cardinality did not respect dimensions: that one-, two-, three-, even n-dimensional sets all had the same cardinality, he said, "I see it, but I don't believe it."

Did Cantor's comment suggest that he found it hard to believe his own theorem even after he had proved it? Byers was by no means the first to say so.

Many mathematicians thinking about the experience of doing mathematics have found Cantor's phrase useful. In his preface to the original (1937) publication of the Cantor-Dedekind correspondence, J. Cavaillès [14, p 3, my translation] already called attention to the phrase:

> . . . these astonishing discoveries—astonishing first of all to the author himself: "I see it but I don't believe it at all,"[1] he writes in 1877 to Dedekind about one of them—, these radically new notions . . .

Notice, however, that Cavaillès is still focused on the description of the result as "surprising" rather than on the issue of Cantor's psychology. It was probably Jacques Hadamard who first connected the phrase to the question of how mathematicians think, and so in particular to what Cantor was thinking. In his famous *Essay on the Psychology of Invention in the Mathematical Field*, first published in 1945 [**10, pp. 61–62**] (only eight years after [**14**]), Hadamard is arguing about Newton's ideas:

> . . . if, strictly speaking, there could remain a doubt as to Newton's example, others are completely beyond doubt. For instance, it is certain that Georg Cantor could not have foreseen a result of which he himself says "I see it, but I do not believe it.".

Alas, when it comes to history, few things are "certain."

The Main Characters

Our story plays out in the correspondence between Richard Dedekind and Georg Cantor during the 1870s. It is important to know something about each of them.

Richard Dedekind was born in Brunswick on October 6, 1831, and died in the same town, now part of Germany, on February 12, 1916. He studied at the University of Göttingen, where he was a contemporary and friend of Bernhard Riemann and where he heard Gauss lecture shortly before the old man's death. After Gauss died, Lejeune Dirichlet came to Göttingen and became Dedekind's friend and mentor.

Dedekind was a creative mathematician, but he was not particularly ambitious. He taught in Göttingen and in Zurich for a while, but in 1862 he returned to his hometown. There he taught at the local Polytechnikum, a provincial technical university. He lived with his brother and sister and seemed uninterested in offers to move to more prestigious institutions. See [1] for more on Dedekind's life and work.

Our story begins in 1872. The first version of Dedekind's ideal theory had appeared as Supplement X to Dirichlet's *Lectures in Number Theory* (based on actual lectures by Dirichlet but entirely written by Dedekind). Also just published was one of his best known works, "Stetigkeit und irrationale Zahlen" ("Continuity and Irrational Numbers"; see [7]; an English translation is included in [5]). This work was his account of how to construct the real numbers as "cuts." He had worked out the idea in 1858 but published it only 14 years later.

Georg Cantor was born in St. Petersburg, Russia, on March 3, 1845. He died in Halle, Germany, on January 6, 1918. He studied at the University of Berlin, where the mathematics department, led by Karl Weierstrass, Ernst Eduard Kummer, and Leopold Kronecker, might well have been the best in the world. His doctoral thesis was on the number theory of quadratic forms.

In 1869, Cantor moved to the University of Halle and shifted his interests to the study of the convergence of trigonometric series. Very much under Weierstrass's influence, he too introduced a way to construct the real numbers, using what he called "fundamental series." (We call them "Cauchy sequences.") His paper on this construction also appeared in 1872.

Cantor's lifelong dream seems to have been to return to Berlin as a professor, but it never happened. He rose through the ranks in Halle, becoming a full professor in 1879 and staying there until his death. See [13] for a short account of Cantor's life. The standard account of Cantor's mathematical work is [4].

Cantor is best known, of course, for the creation of set theory, and in particular for his theory of transfinite cardinals and ordinals. When our story begins, this work was mostly still in the future. In fact, the birth of several of these ideas can be observed in the correspondence with Dedekind. This correspondence was first published in [14]; we quote it from the English translation by William Ewald in [8, pp 843–878].

"Allow Me to Put a Question to You"

Dedekind and Cantor met in Switzerland when they were both on vacation there. Cantor had sent Dedekind a copy of the paper containing his construction of the real numbers. Dedekind responded, of course, by sending Cantor a copy of his booklet. And so begins the story.

Cantor was 27 years old and very much a beginner, while Dedekind was 41 and at the height of his powers; this difference accounts for the tone of deference on Cantor's side of the correspondence. Cantor's first letter acknowledged receipt of [7] and says that "my conception [of the real numbers] agrees entirely with yours," the only difference being in the actual construction. But on November 29, 1873, Cantor moves on to new ideas:

Allow me to put a question to you. It has a certain theoretical interest for me, but I cannot answer it myself; perhaps you can, and
would be so good as to write me about it. It is as follows.

Take the totality of all positive whole-numbered individuals n
and denote it by (n). And imagine, say, the totality of all positive
real numerical quantities x and designate it by (x). The question is
simply, Can (n) be correlated to (x) in such a way that to each individual of the one totality there corresponds one and only one of
the other? At first glance one says to oneself no, it is not possible,
for (n) consists of discrete parts while (x) forms a continuum. But
nothing is gained by this objection, and although I incline to the
view that (n) and (x) permit no one-to-one correlation, I cannot
find the explanation which I seek; perhaps it is very easy.

In the next few lines, Cantor points out that the question is not as dumb
as it looks, since "the totality $\left(\frac{p}{q}\right)$ of all positive rational numbers" *can* be
put in one-to-one correspondence with the integers.

We do not have Dedekind's side of the correspondence, but his notes
indicate that he responded indicating that (1) he could not answer the
question either, (2) he could show that the set of all *algebraic* numbers is
countable, and (3) he didn't think the question was all that interesting.
Cantor responded on December 2:

I was exceptionally pleased to receive your answer to my last letter. I put my question to you because I had wondered about it
already several years ago, and was never certain whether the difficulty I found was subjective or whether it was inherent in the
subject. Since you write that you too are unable to answer it, I
may assume the latter.—In addition, I should like to add that I
have never seriously occupied myself with it, because it has no
special practical interest for me. And I entirely agree with you
when you say that for this reason it does not deserve much effort.
But it would be good if it could be answered; e.g., if it could be
answered with no, then one would have a new proof of Liouville's
theorem that there are transcendental numbers.

Cantor first concedes that perhaps it is not that interesting, then immediately points out an application that was sure to interest Dedekind!
In fact, Dedekind's notes indicate that it worked: "But the opinion I

expressed that the first question did not deserve too much effort was conclusively refuted by Cantor's proof of the existence of transcendental numbers" [8, p 848].

These two letters are fairly typical of the epistolary relationship between the two men: Cantor is deferential but is continually coming up with new ideas, new questions, new proofs; Dedekind's role is to judge the value of the ideas and the correctness of the proofs. The very next letter, from December 7, 1873, contains Cantor's first proof of the uncountability of the real numbers. (It was not the "diagonal" argument; see [4] or [9] for the details.)

"The Same Train of Thought . . ."

Cantor seemed to have a good sense for what question should come next. On January 5, 1874, he posed the problem of higher dimensional sets:

> As for the question with which I have recently occupied myself, it occurs to me that the same train of thought also leads to the following question:
>
> Can a surface (say a square including its boundary) be one-to-one correlated to a line (say a straight line including its endpoints) so that to every point of the surface there corresponds a point of the line, and conversely to every point of the line there corresponds a point of the surface?
>
> It still seems to me at the moment that the answer to this question is very difficult—although here too one is so impelled to say no that one would like to hold the proof to be almost superfluous.

Cantor's letters indicate that he had been asking others about this as well, and that most considered the question just plain weird, because it was "obvious" that sets of different dimensions could not be correlated in this way. Dedekind, however, seems to have ignored this question, and the correspondence went on to other issues. On May 18, 1874, Cantor reminded Dedekind of the question and seems to have received no answer.

The next letter in the correspondence is from May 1877. The correspondence seems to have been reignited by a misunderstanding of what Dedekind meant by "the essence of continuity" in [7]. On June 20,

1877, however, Cantor returns to the question of bijections between sets of different dimensions, and now proposes an answer:

> . . . I should like to know whether you consider an inference-procedure that I use to be arithmetically rigorous.
>
> The problem is to show that surfaces, bodies, indeed even continuous structures of ρ dimensions can be correlated one-to-one with continuous lines, i.e., with structures of only *one* dimension—so that surfaces, bodies, indeed even continuous structures of ρ dimension have the same *power* as curves. This idea seems to conflict with the one that is especially prevalent among the representatives of modern geometry, who speak of simply infinite, doubly, triply, . . . , ρ-fold infinite structures. (Sometimes you even find the idea that the infinity of points of a surface or a body is obtained as it were by squaring or cubing the infinity of points of a line.)

Significantly, Cantor's formulation of the question had changed. Rather than asking *whether* there is a bijection, he posed the question of *finding* a bijection. This is, of course, because he believed he had found one. By this point, then, Cantor knows the right answer. It remains to give a proof that will convince others. He goes on to explain his idea for that proof, working with the ρ-fold product of the unit interval with itself, but for our purposes we can consider only the case $\rho = 2$.

The proof Cantor proposed is essentially this: Take a point (x, y) in $[0, 1] \times [0, 1]$, and write out the decimal expansions of x and y:

$$(x, y) = (0.abcde \ldots , 0.\alpha\beta\gamma\delta\epsilon \ldots).$$

Some real numbers have more than one decimal expansion. In that case, we always choose the expansion that ends in an infinite string of 9s. Cantor's idea is to map (x, y) to the point $z \in [0, 1]$ given by

$$z = 0.a\alpha b\beta c\gamma d\delta e\epsilon \ldots$$

Since we can clearly recover x and y from the decimal expansion of z, this gives the desired correspondence.

Dedekind immediately noticed that there was a problem. On June 22, 1877 (one cannot fail to be impressed with the speed of the German postal service!), he wrote back pointing out a slight problem "which you will perhaps solve without difficulty." He had noticed that the function Cantor had defined, while clearly one-to-one, was not onto. (Of course, he did not use those words.) Specifically, he pointed out that such numbers as

$$z = 0.120101010101\ldots$$

did not correspond to any pair (x, y), because the only possible value for x is $0.100000\ldots$, which is disallowed by Cantor's choice of decimal expansion. He was not sure if this was a big problem, adding "I do not know if my objection goes to the essence of your idea, but I did not want to hold it back."

Of course, the problem Dedekind noticed is real. In fact, there are a great many real numbers not in the image, since we can replace the ones that separate the zeros with *any* sequence of digits. The image of Cantor's map is considerably smaller than the whole interval.

Cantor's first response was a postcard sent the following day. (Can one envision him reading the letter at the post office and immediately dispatching a postcard back?) He acknowledged the error and suggested a solution:

> Alas, you are entirely correct in your objection; but happily it concerns only the proof, not the content. For I proved somewhat more than I had realized, in that I bring a system $x_1, x_2, \ldots, x_\rho$ of unrestricted real variables (that are ≥ 0 and ≤ 1) into one-to-one relationship with a variable y that does not assume all values of that interval, but rather all with the exception of certain y''. However, it assumes each of the corresponding values y' only *once*, and that seems to me to be the essential point. For now I can bring y' into a one-to-one relation with another quantity t that assumes all the values ≥ 0 and ≤ 1.
>
> I am delighted that you have found no other objections. I shall shortly write to you at greater length about this matter.

This is a remarkable response. It suggests that Cantor was very confident that his result was true. This confidence came from the fact that Cantor was already thinking in terms of what later became known as "cardinality." Specifically, he expects that the existence of a one-to-one mapping from one set A to another set B implies that the size of A is in some sense "less than or equal to" that of B.

Cantor's proof shows that the points of the square can be put into bijection with a subset of the interval. Since the interval can clearly be put into bijection with a subset of the square, this strongly suggests that both sets of points "are the same size," or, as Cantor would have said it, "have the same power." All we need is a proof that the "powers" are linearly ordered in a way that is compatible with inclusions.

That the cardinals are indeed ordered in this way is known today as the Schroeder–Bernstein theorem. The postcard shows that Cantor already "knew" that the Schroeder–Bernstein theorem should be true. In fact, he seems to implicitly promise a proof of that very theorem. He was not able to find such a proof, however, then or (as far as I know) ever.

His fuller response, sent two days later on June 25, contained instead a completely different, and much more complicated, proof of the original theorem.

> I sent you a postcard the day before yesterday, in which I acknowledged the gap you discovered in my proof, and at the same time remarked that I am able to fill it. But I cannot repress a certain regret that the subject demands more complicated treatment. However, this probably lies in the nature of the subject, and I must console myself; perhaps it will later turn out that the missing portion of that proof can be settled more simply than is at present in my power. But since I am at the moment concerned above all to persuade you of the correctness of my theorem . . . I allow myself to present another proof of it, which I found even earlier than the other.

Notice that what Cantor is trying to do here is to convince Dedekind that his theorem is true by presenting him a correct proof.[2] There is no indication that Cantor had any doubts about the correctness of the result itself. In fact, as we will see, he says so himself.

Let's give a brief account of Cantor's proof; to avoid circumlocutions, we will express most of it in modern terms. Cantor began by noting that every real number x between 0 and 1 can be expressed as a continued fraction

$$x = \cfrac{1}{a + \cfrac{1}{b + \cfrac{1}{c + \cdots}}}$$

where the partial quotients $a, b, c, \ldots$, etc., are all positive integers. This representation is infinite if and only if x is irrational, and in that case the representation is *unique*.

So one can argue just as before: Given a pair (x, y) such that both x and y are irrational, we "interleave" the continued fractions for x and y to obtain a continued fraction that gives an irrational point in $[0, 1]$. This gives a bijection between the two sets of irrational points. The result is a bijection because the inverse mapping, splitting out two continued fraction expansions from a given one, certainly produces two *infinite* expansions.

That being done, it remains to be shown that the set of irrational numbers between 0 and 1 can be put into bijection with the interval $[0, 1]$. This is the hard part of the proof. Cantor proceeded as follows.

First he chose an enumeration of the rationals $\{r_k\}$ and an increasing sequence of irrationals $\{\eta_k\}$ in $[0, 1]$ converging to 1. He then looked at the bijection from $[0, 1]$ to $[0, 1]$ that is the identity on $[0, 1]$ except for mapping $r_k \mapsto \eta_k, \eta_k \mapsto r_k$. This progression gives a bijection between irrationals in $[0, 1]$ and $[0, 1]$ minus the sequence $\{\eta_k\}$ and reduces the problem to proving that $[0, 1]$ can be put into bijection with $[0, 1] - \{\eta_k\}$.

At this point, Cantor claims that it is now enough to "successively apply" the following theorem:

> A number y that can assume all the values of the interval $(0 \ldots 1)$ with the solitary exception of the value 0 can be correlated one-to-one with a number x that takes on all values of the interval $(0 \ldots 1)$ without exception.

In other words, he claimed that there was a bijection between the half-open interval $(0, 1]$ and the closed interval $[0, 1]$ and that "successive application" of this fact would finish the proof. In the actual application, he would need the intervals to be *open* on the right, so, as we will see, he chose a bijection that mapped 1 to itself.

Cantor did not say exactly what kind of "successive application" he had in mind, but what he says in a later letter suggests it was this: We have the interval $[0, 1]$ minus the sequence of the η_k. We want to "put back in" the η_k, one at a time. So we leave the interval $[0, \eta_1)$ alone and look at (η_1, η_2). Applying the lemma, we construct a bijection between that and $[\eta_1, \eta_2)$. Then we do the same for (η_2, η_3), and so on. Putting together these bijections produces the bijection we want.

Finally, it remained to prove the lemma, that is, to construct the bijection from $[0, 1]$ to $(0, 1]$. Modern mathematicians would probably do this by choosing a sequence x_n in $(0, 1)$, mapping 0 to x_1 and then every

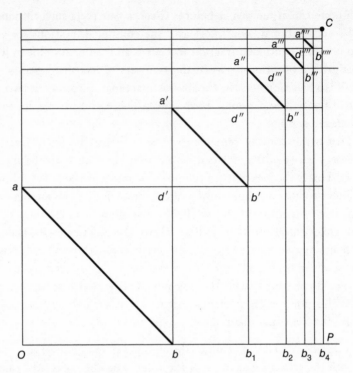

Figure 1. Cantor's function from $[0, 1]$ to $(0, 1]$.

x_n to x_{n+1}. This "Hilbert hotel" idea was still some time in the future, however, even for Cantor. Instead, Cantor chose a bijection that could be represented visually, and simply drew its graph. He asked Dedekind to consider "the following peculiar curve," which we have redrawn in Figure 1 based on the photograph reproduced in [4, p. 63].

Such a picture requires some explanation, and Cantor provided it. The domain has been divided by a geometric progression, so $b = 1/2$, $b_1 = 3/4$, and so on; $a = (0, 1/2)$, $a' = (1/2, 3/4)$, etc. The point C is $(1, 1)$. The points $d' = (1/2, 1/2)$, $d'' = (3/4, 3/4)$, etc., give the corresponding subdivision of the main diagonal.

The curve consists of infinitely many parallel line segments $\overline{ab}$, $\overline{a'b'}$, $\overline{a''b''}$ and of the point c. The endpoints $b, b', b'', \ldots$ are not regarded as belonging to the curve.

The stipulation that the segments are open at their lower endpoints means that 0 is not in the image. This proves the lemma, and therefore the proof is finished.

Cantor did not even add that last comment. As soon as he had explained his curve, he moved on to make extensive comments on the theorem and its implications. He turns on its head the objection that various mathematicians made to his question, namely that it was "obvious" from geometric considerations that the number of variables is invariant:

> For several years I have followed with interest the efforts that have been made, building on Gauss, Riemann, Helmholtz, and others, towards the clarification of all questions concerning the ultimate foundations of geometry. It struck me that all the important investigations in this field proceed from an unproven presupposition which does not appear to me self-evident, but rather to need a justification. I mean the presupposition that a ρ-fold extended continuous manifold needs ρ independent real coordinates for the determination of its elements, and that for a given manifold this number of coordinates can neither be increased nor decreased.
>
> This presupposition became my view as well, and I was almost convinced of its correctness. The only difference between my standpoint and all the others was that I regarded that presupposition as a theorem which stood in great need of a proof; and I refined my standpoint into a question that I presented to several colleagues, in particular at the Gauss Jubilee in Göttingen. The question was the following:
>
> "Can a continuous structure of ρ dimensions, where $\rho > 1$, be related one-to-one with a continuous structure of one dimension so that to each point of the former there corresponds one and only one point of the latter?"
>
> Most of those to whom I presented this question were extremely puzzled that I should ask it, for it is quite *self-evident* that the determination of a point in an extension of ρ dimensions always needs ρ independent coordinates. But whoever penetrated the sense of the question had to acknowledge that a proof was needed to show why the question should be answered with the "self-evident" *no*. As I say, I myself was one of those who held it for the *most likely* that the question should be answered with

a *no*—until quite recently I arrived by rather intricate trains of thought at the conviction that the answer to that question is an unqualified *yes*.[3] Soon thereafter I found the proof which you see before you today.

So one sees what wonderful power lies in the ordinary real and irrational numbers, that one is able to use them to determine uniquely the elements of a ρ-fold extended continuous manifold *with a single coordinate*. I will only add at once that their power goes yet further, in that, as will not escape you, my proof can be extended without any great increase in difficulty to manifolds with an infinitely great dimension-number, provided that their infinitely-many dimensions have the form of a simple infinite sequence.

Now it seems to me that all philosophical or mathematical deductions that use that erroneous presupposition are inadmissible. Rather the difference that obtains between structures of *different* dimension-number must be sought in quite other terms than in the number of independent coordinates—the number that was hitherto held to be characteristic.

"Je Le Vois . . ."

So now Dedekind had a lot to digest. The interleaving argument is not problematic in this case, and the existence of a bijection between the rationals and the increasing sequence η_k had been established in 1872. But there were at least two sticky points in Cantor's letter.

First, there is the matter of what kind of "successive application" of the lemma Cantor had in mind. Whatever it was, it would seem to involve constructing a bijection by "putting together" an infinite number of functions. One can easily get in trouble.

For example, here is an alternative reading of what Cantor had in mind. Instead of applying the lemma to the interval (η_1, η_2), we could apply it to $(0, \eta_1)$ to put it into bijection with $(0, \eta_1]$. So now we have "put η_1 back in," and we have a bijection between $[0, 1] - \{\eta_1, \eta_2, \eta_3, \ldots\}$ and $[0, 1] - \{\eta_2, \eta_3, \ldots\}$.

Now repeat: Use the lemma on $(0, \eta_2)$ to make a bijection to $(0, \eta_2]$. So we have "put η_2 back in." If we keep doing that, we presumably get a bijection from $(0, 1)$ minus the η_k to all of $(0, 1)$.

But do we? What is the image of, say, $\frac{1}{3}\eta_1$? It is not fixed under any of our functions. To determine its image in [0, 1], we would need to compose infinitely many functions, and it's not clear how to do that. If we manage to do it with some kind of limiting process, then it is no longer clear that the overall function is a bijection.

The interpretation Cantor probably intended (and later stated explicitly) yields a workable argument because the domains of the functions are disjoint, so it is clear where to map any given point. But since Cantor did not indicate his argument in this letter, one can imagine Dedekind hesitating. In any case, at this point in history the idea of constructing a function out of infinitely many pieces would have been both new and worrying.

The second sticky point was Cantor's "application" of his theorem to undermine the foundations of geometry. This is, of course, the sort of thing one has to be careful about. And it is clear, from Dedekind's eventual response to Cantor, that it concerned him.

Dedekind took longer than usual to respond. Having already given one wrong proof, Cantor was anxious to hear a "yes" from Dedekind, and so he wrote again on June 29:

> Please excuse my zeal for the subject if I make so many demands upon your kindness and patience; the communications which I lately sent you are even for me so unexpected, so new, that I can have no peace of mind until I obtain from you, honoured friend, a decision about their correctness. So long as you have not agreed with me, I can only say: *je le vois, mais je ne le crois pas.* And so I ask you to send me a postcard and let me know when you expect to have examined the matter, and whether I can count on an answer to my quite demanding request.

So here is the phrase. The letter is, of course, in German, but the famous "I see it, but I don't believe it" is in French.[4] Seen in its context, the issue is clearly not that Cantor was finding it hard to believe his *result*. He was confident enough about that to think he had rocked the foundations of the geometry of manifolds. Rather, he felt a need for confirmation that his *proof* was correct. It was his *argument* that he saw but had trouble believing. This issue is confirmed by the rest of the letter, in which Cantor spelled out in detail the most troublesome step, namely, how to "successively apply" his lemma to construct the final bijection.

So the famous phrase does not really provide an example of a mathematician having trouble believing a theorem even though he had proved it. Cantor, in fact, seems to have been confident [*überzeugt!*] that his theorem was true, as he himself says. He had in hand at least two arguments for it: The first argument, using the decimal expansion, required supplementation by a proof of the Schroeder–Bernstein theorem, but Cantor was quite sure that this would eventually be proved. The second argument was correct, he thought, but its complicated structure might have allowed something to slip by him.

He knew that his theorem was a radically new and surprising result—it would certainly surprise others!—and thus it was necessary that the proof be as solid as possible. The earlier error had given Cantor reason to worry about the correctness of his argument, leaving Cantor in need of his friend's confirmation before he would trust the proof.

Cantor was, in fact, in a position much like that of a student who has proposed an argument, but who knows that a proof is an argument that convinces his teacher. Though no longer a student, he knows that a proof is an argument that will convince others and that in Dedekind he had the perfect person to find an error if one were there. So he saw, but until his friend's confirmation he did not believe.

What Came Next

So why did Dedekind take so long to reply? From the evidence of his next letter, dated July 2, it was not because he had difficulty with the proof. His concern, rather, was Cantor's challenge to the foundations of geometry.

The letter opens with a sentence clearly intended to allay Cantor's fears: "I have examined your proof once more, and I have discovered no gap in it; I am quite certain that your interesting theorem is correct, and I congratulate you on it." But Dedekind did not accept the consequences Cantor seemed to find:

> However, as I already indicated in the postcard, I should like to make a remark that counts *against* the conclusions concerning the concept of a manifold of ρ dimensions that you append in your letter of 25 June to the communication and the proof of the theorem. Your words make it appear—my interpretation may be

incorrect—as though on the basis of your theorem you wish to cast doubt on the meaning or the importance of this concept . . .

Against this, I declare (despite your theorem, or rather in consequence of reflections that it stimulated) my conviction or my faith (I have not yet had time even to make an attempt at a proof) that the dimension-number of a continuous manifold remains its first and most important invariant, and I must defend all previous writers on the subject. . . . For all authors have clearly made the tacit, completely natural presupposition that in a new determination of the points of a continuous manifold by new coordinates, these coordinates should also (in general) be *continuous* functions of the old coordinates . . .

Dedekind pointed out that, in order to establish his correspondence, Cantor had been "compelled to admit a frightful, dizzying discontinuity in the correspondence, which dissolves everything to atoms, so that every continuously connected part of one domain appears in its image as thoroughly decomposed and discontinuous." He then set out a new conjecture that spawned a whole research program:

. . . for the time being I believe the following theorem: "If it is possible to establish a reciprocal, one-to-one, and complete correspondence between the points of a continuous manifold A of *a* dimensions and the points of a continuous manifold B of *b* dimensions, then this *correspondence itself*, if *a* and *b* are unequal, is necessarily *utterly discontinuous*."

In his next letter, Cantor claimed that this was indeed his point: Where Riemann and others had casually spoken of a space that requires *n* coordinates as if that number was known to be invariant, he felt that this invariance required proof. "Far from wishing to turn my result against the article of faith of the theory of manifolds, I rather wish to use it to secure its theorems," he wrote. The required theorem turned out to be true, indeed, but proving it took much longer than either Cantor or Dedekind could have guessed: It was finally proved by Brouwer in 1910. The long and convoluted story of that proof can be found in [3], [11], and [12].

Finally, one should point out that it was only some three months later that Cantor found what most modern mathematicians consider the "obvious" way to prove that there is a bijection between the interval minus a countable set and the whole interval. In a letter dated October 23, 1877, he took an enumeration ϕ_ν of the rationals and let $\eta_\nu = \sqrt{2}/2^\nu$. Then he constructed a map from $[0, 1]$ sending η_ν to $\eta_{2\nu-1}$, ϕ_ν to $\eta_{2\nu}$, and every other point h to itself, thus getting a bijection between $[0, 1]$ and the irrational numbers between 0 and 1.

Mathematics as Conversation

Is the real story more interesting than the story of Cantor's surprise? Perhaps it is, since it highlights the social dynamic that underlies mathematical work. It does not render the theorem any less surprising but shifts the focus from the result itself to its proof.

The record of the extended mathematical conversation between Cantor and Dedekind reminds us of the importance of such interaction in the development of mathematics. A mathematical proof is, after all, a kind of challenge thrown at an idealized opponent, a skeptical adversary who is reluctant to be convinced. Often, this adversary is actually a colleague or collaborator, the first reader and first critic.

A proof is not a proof until some reader, preferably a competent one, says it is. Until then we may see, but we should not believe.

Notes

1. Cavaillès misquotes Cantor's phrase as "je le vois mais je ne le crois point."

2. Cantor claimed that he had found this proof before the other. I find this notion hard to believe. In fact, the proof looks much like the result of trying to fix the problem in the first proof by replacing (nonunique) decimal expansions with (unique) continued fraction expansions.

3. The original reads ". . . bis ich vor ganz kurzer Zeit durch ziemlich verwickelte Gedankereihen zu der Ueberzeugung gelangte, dass jene Frage ohne all Einschränkgung zu bejahen ist." Note Cantor's *Überzeugung*—conviction, belief, certainty.

4. I don't know whether this is because of the rhyme vois/crois, or because of the well-known phrase "voir, c'est croire," or for some other reason. I do not believe the phrase was already proverbial.

References

1. K.-R. Biermann, Dedekind, in *Dictionary of Scientific Biography*, C. C. Gillispie, ed., Scribners, New York, 1970–1981.

2. W. Byers, *How Mathematicians Think: Using Ambiguity, Contradiction, and Paradox to Create Mathematics*, Princeton University Press, Princeton, NJ, 2007.

3. J. W. Dauben, "The invariance of dimension: Problems in the early development of set theory and topology," *Historia Math.* **2** (1975) 273–288. doi:10.1016/0316-0860(75) 90066-X.

4. ———, *Georg Cantor: His Mathematics and Philosophy of the Infinite*, Princeton University Press, Princeton, NJ, 1990.

5. R. Dedekind, *Essays in the Theory of Numbers* (trans. W. W. Beman), Dover, Mineola, NY, 1963.

6. ———, *Gesammelte Mathematische Werke*, R. Fricke, E. Noether, and O. Ore, eds., Chelsea, New York, 1969.

7. ———, *Stetigkeit and Irrationalzahlen und irrationale Zahlen*, 1872, in *Gesammelte Mathematische Werke*, vol. 3, item L, R. Fricke, E. Noether, and O. Ore, eds., Chelsea, New York, 1969.

8. W. Ewald, *From Kant to Hilbert: A Source Book in the Foundations of Mathematics*, Oxford University Press, Oxford, U.K., 1996.

9. R. Gray, "Georg Cantor and transcendental numbers," *Amer. Math. Monthly* **101** (1994) 819–832. doi:10.2307/2975129.

10. J. Hadamard, *An Essay on the Psychology of Invention in the Mathematical Field*, Princeton University Press, Princeton, NJ, 1945.

11. D. M. Johnson, "The problem of the invariance of dimension in the growth of modern topology I," *Arch. Hist. Exact Sci.* **20** (1979) 97–188. doi:10.1007/BF00327627.

12. ———, "The problem of the invariance of dimension in the growth of modern topology II," *Arch. Hist. Exact Sci.* **25** (1981) 85–267. doi:10.1007/BF02116242.

13. H. Meschkowski. Cantor, in *Dictionary of Scientific Biography*, C. C. Gillispie, ed., Scribners, New York, 1970–1981.

14. E. Noether and J. Cavailles, *Briefwechsel Cantor-Dedekind*, Hermann, Paris, 1937.

Why Is There Philosophy of Mathematics at All?

IAN HACKING

A Perennial Philosophical Obsession

Mathematics is the only specialist branch of human knowledge that has consistently obsessed many dead great men in the Western philosophical canon. Not all, for sure, but Plato, Descartes, Leibniz, Kant, Husserl, and Wittgenstein form a daunting array. And that list omits the angry skeptics about the significance of mathematical knowledge, such as Berkeley and Mill, and the logicians, such as Aristotle and Russell.

Why has mathematics mattered to so many famous philosophers? And why does it infect, in many cases, their entire philosophies? Aside from the naysayers, such as Mill, it is first of all because they have *experienced* mathematics and found it passing strange. The mathematics that they have encountered has *felt* different from other experiences of observing, learning, discovering, or simply "finding out." This difference is partly because the gold standard for finding out in mathematics is demonstrative proof. Not, of course, any old proof, for the most boring things can be proven in the most boring ways. I mean proofs that deploy new ideas in unexpected ways, proofs that can be understood, proofs that embody ideas that are pregnant with further developments. Mathematicians still like to cite Euclid's proof that there are infinitely many primes, or the proof that the square root of 2 is not a rational fraction.

Most people do not respond to mathematics with such experiences or feelings; they really have no idea what is moving those philosophers. They are in good company. Take Hume, one of my heroes, who can do no wrong. He was one of the most brilliant reasoners who trod the face of the earth, but it is quite possible that he was never especially impressed by deductive proofs. "Experiencing mathematics" in no way implies the possession of philosophical gifts—perhaps the opposite.

Indeed one strange thing about the philosophers' fascination with mathematics is that most people pay it no heed at all. For the majority, mathematics is something hateful that is imposed by teachers and is to be escaped as quickly as possible. Today's cognitive science is plausible when it maintains that there are "innate" cognitive modules of arithmetic and geometry that have evolved, perhaps around the time that linguistic modules evolved (e.g., Dehaene 1997, Butterworth 1999, Spelke et al. 2010). But this must be put beside the fact that whereas a high degree of linguistic competence is universally acquired early in life, even modest mathematical competence beyond rote learning is acquired by only a small proportion of the population. To emphasize this contrast, recall Chomsky's insistence on the creative aspect of language use. He claimed that this is a human universal. Then notice that the capacity for even modestly creative uses of mathematics is not widely shared among humans, despite our forcing the basics on the young. So the philosophical obsession is all the more puzzling.

PROOF

The human capacity for finding out new facts by deductive proofs was discovered very late in the history of the human race, at a particular time and place. Kant exaggerated, in the *preface* to the second edition of the *Critique of Pure Reason* (1787), when he said that the discovery was the result of "the happy thought of a single man," "be he Thales or some other" (Kant 1930, 19–20, Bix–xii).[1] But he was right to say that "The transformations must have been due to a *revolution*" (his emphasis). In the same paragraph, he calls it an "intellectual revolution." And he uses the word a third time in the same paragraph. He was indeed wrong to say that the revolution was effected by "the happy thought of a single man." But something radical happened.

Reviel Netz puts the revolution later than Kant does, about the time of Eudoxus or a little earlier. Using a metaphor favoured in paleontology, Netz says that "the early history of Greek mathematics was catastrophic. . . . A relatively large number of interesting results would have been discovered practically simultaneously" (Netz 1999, 273). The human capacity for deductive proof seems to have been discovered by a few individuals who built a small and esoteric community around it—mostly by correspondence, and by founding a few small schools. There are interesting speculations as to why there was "uptake" only in Greece, but that takes us too far afield (Lloyd 1990 and Netz 1999, 205ff.).

I do not mean to suggest that Greek mathematics was an "immaculate conception" (to use an apt expression used in correspondence with Jens Høyrup). The people who discovered proof were familiar with a rich trove of mathematical practices in Mesopotamia and North Africa. But there was no philosophy of mathematics until there was the experience of proving something demonstratively. That was one of the so-called Greek miracles.

ANCIENT AND ENLIGHTENMENT

I now suggest that the reasons why there is philosophy of mathematics divide roughly in two. The experience of proof figures in both, but in different ways. For convenience I label one Ancient and the other Enlightenment. The emblem of the Ancient, here, is Plato. The emblem of the Enlightenment, here, is Kant. The two strands interweave promiscuously, but it is helpful to distinguish them.

The Enlightenment strand is more impressive to philosophers than to mathematicians. The Ancient strand is the one that appears most strongly when mathematicians express philosophical views. I shall say nothing new about famous issues in the philosophy of mathematics. But the classification proposed here may rearrange some of their relationships.

An Ancient Strand in the Answer: "Out There"

To speak of an Ancient strand is not to be stuck in the past; I am discussing perennial concerns. So I shall use a debate in which a self declared "Platonist" or "realist" mathematician faced off against a "nominalist" neurobiologist. The two eminent protagonists were Alain Connes and Jean-Pierre Changeux (1989, 1995).[2] Connes cannot doubt that there is a mathematical reality "out there," independent of human thought. "The working mathematician can be likened to an explorer who discovers the world" (p 12). This is the same metaphor used by G. H. Hardy in his well-known essay, "Mathematical Proof"

(1929). Notice that Hardy, despite his title, emphasized exploration rather than formal proof.

Changeux speaks of Connes' Platonism. Connes protests: "The term 'realism,' by the way, is to be preferred to your misleading 'platonism'" (p. 31). Unlike Connes, I prefer "Platonism" over "realism" just because in this context it is usually restricted to mathematics, whereas "realism" runs not only to "scientific realism" and the entire gamut of great medieval thinkers, and also, for example, to Zola's novels.

Names don't matter: We know Connes' countryside, however it be named. The question is, why should his attitude be so perennial, and also so perennially challenged?

Changeux's challenge is recognizably a form of antirealism about mathematics, but of an interesting sort that only recently could be grounded on empirical knowledge. He is a neurobiologist with access to the resources of brain science and the cognitive sciences. He is convinced that mathematical structures are by-products of the innate endowments of the human brain. He goes so far as to say that "mathematical objects exist materially in the brain" (p. 13).

THE MONSTER

I said that the Ancient source of philosophy of mathematics lies in the experience of proof and more generally mathematical discovery. Connes elegantly illustrates this thesis—not only why he is a "Platonist," but also why there is a philosophy at all.

Connes (p. 19) is enormously impressed with a fact about mathematical research. His example is thoroughly up to date yet is of just the same form as examples that could have been used in classical Athens.

Here we come upon a characteristic peculiar to mathematics that is very difficult to explain. Often it's possible, although only after considerable effort, to compile a list of mathematical objects defined by very simple conditions. Intuitively one believes that the list is complete, and searches for a general proof of its exhaustiveness. New objects are frequently discovered in just this way, as a result of trying to show that the list is exhausted. Take the example of finite groups. The notion of a finite group is elementary, almost on the same level as that of an integer. A finite group is the group of symmetries of a finite object. Mathematicians have struggled to classify the finite simple groups, that is to say the finite groups that (like the prime numbers to some extent) can't be decomposed into smaller groups. . . . About fifteen years [viz. about 1974[3]] ago the last finite simple group—the 'Monster'— was discovered by purely mathematical reasoning. It is a finite group with a considerable number of elements: 808,017,424,794, 512,875,886,459,904,961,710,757,005,754,368,000,000,000

It has now at last been shown, as a result of heroic efforts, that the list of twenty-six finite simple groups is indeed complete.

Connes writes as if the monster has been sitting out there, quietly grinning, waiting for us to discover it.

Mathematicians soon made some sense of the monster. That began with the prolifically creative mathematician John Conway—who had named the monster in the first place—and what he called the Monstrous Moonshine Conjecture. The idea seemed so preposterous that Conway's first reaction was "Moonshine!" The monster turned out to be identical to an object derived from a completely different branch of mathematics. It *had* to be a coincidence! Except it was instead one of those familiar cases of underlying unity within the diversity of mathematics. Here I want only to emphasize Connes' heartfelt *feeling* or *experience*, that this at first sight absurd object was *just there*, waiting for us. And a little later, this monstrous object turned out to be the very same as one identified in a completely different field of mathematics. Out there? We are reminded of Frege's example, in which the morning star turned out to be identical to the evening star.

Connes' reaction is quite a common one. Richard Borcherds won the Fields Medal in 1998 for proving the moonshine conjecture. In conversation he said, "When you think about the Monster you have to wonder who *made* it! It is almost like that Intelligent Design stuff; the Monster has such a complex and yet organized structure, that it is as if it had been engineered by someone."[4]

GLITTER

Borcherds was honestly expressing his persistent astonishment about the sheer existence of the monster. It is that kind of astonishment that engenders philosophy of mathematics, and therein lies the Ancient strand in the very existence of perennial issues.

Borcherds does not imagine that the monster had a designer; he said only that the object is so delicate in all its parts that it is as if it had been designed. He was not advancing an opinion. He was expressing heart-felt incredulity at a fact that, in any ordinary sense, he understood at least as well as anyone else in the world. He was not so much surprised at the fact as by the *existence* of such facts.

Connes was giving vent to exactly the same sentiment. But he was using it to advance his own philosophy of mathematics. One may com-plain that he was also using a certain sort of mathematician's rhetoric to scare us into submission. (It did not work on Changeux, who finds the monster a bore.) We are reminded of Wittgenstein's talk of *glit-ter* (1978, 274). Wittgenstein was probably referring, in this context, to Cantor's transfinite, but the metaphor can be applied to Connes' rhetoric. We are supposed to be impressed when he writes down this meaningless (to us) number, but somehow it is overdone; instead, we're offered what Wittgenstein called "*mysteriousness*" (1978, original emphasis).

"All that I can do [Wittgenstein continued] is to shew an easy escape from this obscurity and this glitter of the concepts." I do not believe there is "an easy escape from this obscurity and glitter of the concepts" illustrated by Connes. It just is astonishing, that an elementary charac-terization of groups should generate this strange "monster."

ARCHAIC MATHEMATICAL REALITY

Connes' philosophizing about mathematics arises from what I call an Ancient (classically Greek) experience of mathematics. It is now time to say that his own philosophy is not that of an indiscriminate "Platonist." He does not think that all mathematical objects are like the monster, grinning and waiting to be found. He agrees that most of the tools devised by mathematicians are inventions, not discoveries. His label for such tools is "projective." But they are used to investigate what he calls archaic (or primordial) *mathematical reality*[5] (Connes and Changeux 1995, 192).

This is his way of expressing a realism about mathematics that is both modest and specific. We construct mathematical tools in abundance, he writes, but what is remarkable is that using them we can identify uniquely various objects that are not, in his opinion, constructed. They constitute archaic, that is, primitive, original, mathematical reality. The integers are, in his judgment, a familiar part of that reality. We are reminded of the aphorism attributed to Leopold Kronecker (1823–1891): "God made the integers and all else is the work of man." (Cantor was Kronecker's student: Is this an oedipal reaction?) I do not mean to imply that Connes is anti-Cantorian, only that the spirit of his philosophy is reminiscent of Kronecker.

I do not share Connes' convictions, but I find them more instructive than blanket Platonism, which says, without discrimination, that "abstract objects exist" or, following Quine, "anything over which we quantify exists."

NIETZSCHEAN DEFLATION

One can derive from Nietzsche the suggestion that mathematical objects are by-products of grammar—more explicitly, of our practice of expressing mathematics in propositions—and the fact that European languages demand an existential presupposition for terms in the subject position of a declarative sentence. Perhaps Wittgenstein had this in mind when he said (1978, 274),

> Is it already mathematical alchemy, that mathematical propositions are regarded as statements about mathematical objects,— and mathematics as the exploration of those objects?

The Nietzschean use of grammar to undermine Platonism is powerful against its blanket formulations, but, perhaps, not against Connes' limited form. Connes may, like all of us, be sculpted by grammar, but he is moved by the experience of exploring a specific "reality" that is "out there"—not any old abstract objects, but the primordial ones.

THE NEUROBIOLOGICAL RETORT

Jean-Pierre Changeux holds that mathematical truth is constrained by the neuronal structure of the brain. In answer to Connes, he retorts that we have here only a complicated version of the finite list of regular polyhedra, a list that so impressed Plato that they are called the platonic solids. They come at the end of the last book of Euclid (XIII), plus a proof that there are only five of them. It has even been proposed that the entire point of the *Elements* was to reach this peak of discovery. So Changeux's taunt, "Same old story" backfires. We can imagine Plato having used the regular polyhedra in an argument identical to that of Alain Connes, but with the classification of the regular polyhedra in place of the classification of the finite simple groups. Indeed a popular exposition of the monster and its confrères by Mark Ronan (2006) actually begins with the platonic solids. Connes did choose something new and "glittering" precisely because we have become blase about polyhedra, but the point is exactly the same.

Here we get a vindication of my description of this strand in the persistence of philosophy of mathematics, "Ancient." The passage from the platonic solids to the monster is well trodden and is as broad and attractive to many mathematicians today, as it was in Ancient times.

And, just as in ancient times, this route is not attractive to all mathematicians. A notable counterexample is Timothy Gowers (Fields Medal 1998). Gowers (2006) explicitly rejects Connes' Platonism, although he does not address the idea of "archaic reality." It is relevant that Gowers is the only major active mathematician known to me who acknowledges Wittgenstein's effect on his own thinking. "Anyone who has read this book [the wonderful *Mathematics: A Very Short Introduction*] and the *Philosophical Investigations* will see how much the later Wittgenstein has influenced my philosophical outlook and in particular my views on the abstract method" (Gowers 2002, 140).

MY ATTITUDE TO THIS DEBATE

The debate between Connes and Changeux wears contemporary garb, but its form is perennial. Hence I am not going to settle anything. It would, however, be disingenuous not to declare my sympathies. My own opinion is much more obscure than that of either of the two controversialists. I think that what Connes says is right but what he means is wrong. He really does believe in a fabulous domain of numbers, archaic and primordial, whose structures have nothing to do with the brain. That *is* (I assert) what he means, and it is in my opinion wrong. I think that what Changeux says is wrong—"mathematical objects exist materially in the brain"—but what he means is right. He means that the structures

Connes so admires are by-products of our genetic envelope (a phrase I got from Changeux himself, but which he seems not to use in print).

NEW BUT SCHOLASTIC "PLATONISM"

Examples of the Ancient have, unsurprisingly, centred on Platonism. There are endless arguments pro and con Platonism about math, including new ones that have nothing to do with mathematical experience, or the intense feeling that mathematics is just "out there." I call two of the most influential recent examples *scholastic* not to demean their importance to philosophy but because mathematicians and the larger intellectual worlds have little interest in them.

One prime example is Quine's indispensability argument. Starting with the idea that "to be is to be the value of a variable" and the fact that mathematics is used in all branches of science, he argued that we have to admit that numbers exist. Hilary Putnam went so far as to say that it would be intellectually dishonest not to do so. So some sort of Platonism must be true. Hartry Field tried to refute that by showing that we could do physics without numbers. (For references, see Colyvan 1998.) This is an interesting debate, but only to philosophers. Mark Balaguer (1998) concludes his survey of all possible debates about mathematical Platonism by saying that the indispensability argument is the best of available arguments (none of which is conclusive). Yet it has never moved a mathematician or a member of the republic of letters, outside the narrow confines of academic philosophy, to budge in favor of Platonism.

A second example is Benacerraf's dilemma (1973). Platonists can explain what mathematical truths are about, using standard referential semantics. But they cannot explain how we find out about abstract objects, which are by definition not accessible to any kind of sense experience or any causal relation between our knowledge and what we know about. Constructivists can explain how we know: by proof! But what is it that we are knowing about? This is such a dilemma for philosophers that some take Benacerraf to favor Platonism while others take him to be opposed to it! Once again, this is a debate among philosophers and logicians, which has few echoes outside their community.

Benacerraf's dilemma does, however, lead us on to the second strand, Enlightenment, in reasons for there being philosophy of mathematics. Michael Dummett construed what is essentially one horn of the dilemma in this way: "How can we know anything about this [Platonist] realm of immaterial objects? And how can facts about it have any relevance to the physical universe we now inhabit?—how, in other words, could a mathematical theory, so understood, be *applied*?" (Dummett 1991, 301). The "applied" is at the core of the Enlightenment strand, but it works at a number of different levels because the idea of applying mathematics can be taken in a number of different ways, each with its own roster of philosophical interests.

An Enlightenment Strand in the Answer: Necessary Truth, a Priori Knowledge, and Application

RUSSELL'S ANSWER TO "WHY?"

When he had finished and done with *Principia*, Bertrand Russell sat down to write a potboiler that has charmed young people ever since: *The Problems of Philosophy* (1912). It charms me still. In the course of covering the waterfront, he wrote that "every philosophy which is not purely sceptical" must find an answer to Kant's question, "How is pure mathematics possible?" Russell exaggerated. Some great philosophies that are not purely skeptical have had no interest in mathematics whatsoever. But the question stands. What is so strange about mathematics, that it should tie so many philosophers into knots? Russell mentioned one source of wonder: "The apparent power of anticipating facts about things of which we have no experience is certainly surprising" (1912, 85).

Later in the book, using the example of two plus two makes four, Russell argues that the surprise is unwarranted. This is because "the statement 'two and two are four' deals exclusively with universals, and therefore may be known by anybody who is acquainted with the universals and can perceive the relation between them." And "it must be taken as a fact" that we can do this. Leaving that aside, we can, with generosity, read the rest of his paragraph as a succinct response to Kant's question. "Although our general proposition is *a priori*, all its applications to actual particulars involve experience and therefore contain an empirical element" (1912, 105).

Russell's response is plainly connected with Frege's theory of number—a theory that, I suspect, we would not even know about had not Russell rediscovered it. (It was first presented for a wide audience in Russell's 1903 *Principles of Mathematics*.) Where Russell speaks vulgarly of universals, Frege spoke of concepts, so that, in overly simple shorthand, to say that there five pencils on this table is to say something about the concept "pencil on this table." Mark Steiner (1998, 16–23) has a rigorous analysis of how Frege's theory should be spelled out. (It thereby provides a way to understand Russell's hand-waving "acquaintance" with universals.) Steiner then goes on to discuss issues about the applicability of mathematics that Frege did not resolve.

KANT'S ARGOT

It is still worth recalling Kant's glorious exclamation of awe, the one that put together our philosophical argot. The *Prolegomena to any future Metaphysics that Will be Able to Come Forward as Science* (1783) comes after the first critique (1781) but before the "historicist" preface to the second (1787).

> Here is a great and proven body of knowledge,[6] which is already of admirable extent and promises unbounded expansion in the future, which carries with it thoroughly apodictic certainty (i.e. absolute necessity), hence rests on no grounds of experience, and so is a pure product of reason, but beyond this is thoroughly synthetic. 'How is it then possible for human reason to achieve such knowledge wholly *a priori*?' (Kant 1997, 32)

Kant shouted. Not just certainty, but *apodictic* certainty! *Absolute* necessity! *Thoroughly* synthetic! Such shouts try to draw attention to various experiences that some people have, in connection with mathematics. They are connected with the feeling that what is proved must be true. Proofs are compelling; what they establish is certain. We start from proof, as in the Ancient strand, but the Enlightenment led us to focus on the certainty of what is proved and the fact that it somehow has to be true, could not be otherwise.

Add in "a priori" knowledge, or, in Kant, judgment. The a priori is double-edged, for it gets its effect from two aspects. One is the thought that we can find out "just by thinking." The other is that what we find out is true, true of the world. The two together create a terrible conundrum. Kant shouts again, "How is that possible?!!"

APPLICATION

You may feel that Kant was just giving an Enlightenment wording to Ancient worries. Yes, of course, but importantly no. I appeal to Miles Burnyeat's (2000) assertion that Plato had no concept of (logical) necessity. Mathematical truths mattered to him because they are eternal. Logical necessity, as we know it, is an invention of the modern era, derived from concepts of medieval Islam and Christendom, and harking back to very different notions in Aristotle. Of course, it ties in with the Ancient astonishment with proof, for we seem to prove things that must be true, cannot be otherwise. And how can that be?

The wording of Kant's conundrum, "How is pure mathematics possible?" misleads us because words have evolved. Francis Bacon seems to have been the first explicitly to sketch an organization of knowledge with "pure mathematics" and "mixed mathematics" as two of its branches (Brown 1991). "Mixed" was transmuted into "applied," with somewhat different connotations, soon after the second edition of Kant's first *Critique*. And our current distinction between pure and applied mathematics, though foreshadowed in Kant, d'Alembert, and even Bacon, is firmed up only during the nineteenth century (Maddy 2008).

Kant's conundrum (How is pure mathematics possible?) does not derive from what, throughout the twentieth century, was called "pure mathematics." It arose from the *application* of ("pure") mathematics. Thus I have no quarrel with Mark Steiner (1998), who asked more or less my title question, "Why is there philosophy of mathematics?" He answered, in effect, *application*.

The "application" of mathematics covers quite a few things, and Steiner distinguished several. Russell's neat phrase captures the one that probably led Kant to formulate the very concept of the a priori. I mean what Russell called the apparent power of anticipating things of which we have no experience—that is, the power of applying our mathematics to know what's true in the material world around us. But you can't do that unless you know that the relevant mathematical proposition is true, and that commonly comes from its being proved. So we are back to my starting point, proof.

To parody a famous saying of Kant himself: Application without proof would be mere dogma and would often not work. Conversely, proof without some sense of potential application is a mere game. Quine and Wittgenstein agreed on that, for very different reasons.

Kant's Conundrum of a Priori Knowledge Becomes a Twentieth Century Dilemma

Let's have a quick update to one classic twentieth century approach to the conundrum. We can think of it as a dilemma. The Vienna Circle took one horn and Quine the other.

Vienna taught that "$5 + 7 = 12$" is not "a fact about items in the world," although there are many facts, often expressed briefly by the very same sentence, for example, the fact that the five loaves brought by Joseph to the picnic plus the seven brought by Peter make 12 in all. This fact is contingent (Peter is feckless, and he might have eaten one on the way) and is checked empirically if need be. But "$5 + 7 = 12$" is true in virtue of the meanings of the words, analytic. Hempel (1949) is a classic exposition of the connections between the two propositions.

Of course "$5 + 7 = 12$" was, from Kant to now, just a token for myriad mathematical propositions.

Quine demolished the concept of analyticity to his own satisfaction, and that of many others. Hence he could not invoke two propositions, one synthetic and the other analytic. There is only one sentence, "$5 + 7 = 12$." Quine thought that mathematical statements can be called true or false only insofar as they have some real-world application. Totally "pure" mathematics with no application can be called true only as a courtesy. He well knows (Quine 2008, 468) that there is a

> vast proliferation of mathematics that there is no thought or prospect of applying. I see these domains as integral to our overall theory of reality only on sufferance: . . . So it is left to us to try to assess these sentences also as true or false, if we care to. Many are settled by the same laws that settle applicable mathematics. For the rest, I would settle them as far as practicable by considerations of economy, on a par with the decisions we make in natural science when trying to frame empirical hypotheses worthy of experimental testing.

The only mathematical utterances that are properly called true or false are ones that have application, and they are to be assessed as true or false only as part our whole "conceptual scheme."

What about the ideas of necessity? Quine's answer rings clear and loud in many texts. Here's an early one (Quine 1950, xiii):

> Our system of statements has such a thick cushion of indeterminacy, in relation to experience, that vast domains of law can easily be held immune to revision on principle. . . . Mathematics and logic, central as they are to the conceptual scheme, tend to be accorded such immunity, in view of our conservative preference for revisions which disturb the system least; and herein, perhaps, lies the "necessity" which the laws of mathematics and logic are felt to enjoy.

Kant's conundrum has been turned into a dilemma. The Vienna Circle grasps one horn: There are *two* propositions, only one of which is synthetic. Quine grasps the other horn: Necessity is no more than pragmatic immunity. Only mathematics with application is true or false, and, if true, it is true in the same way that anything empirical is true.

These are both great ideas. Each, in its day, has captured a generation of analytic philosophers. Yet one may feel that both horns of the dilemma are too scholastic, too internal to analytic philosophy to have satisfied Kant. We get no explanation of the feeling of conviction, of what I called the awe that prompted Kant's argot in the first place.

APPLICATION

The label, "applied mathematics" leads us to think that the idea of applying mathematics is simple and straightforward. Not so. There are many types of application. I shall offer one classification into types. Steiner does it differently, and there are many other possible distinctions to be made. I distinguish six different types of application that for brevity I shall call *Apps*. Very roughly, they go from what is in one sense, fundamental (*App* 0) to what many people would regard as bizarre (*App* 6). I should warn that this seemingly vertical tower of *Apps* is pretty wobbly. It is not a tower of chronological or intellectual development.

Philosophers, physicists, and journalists discussing the application of mathematics often think of the exclamations of surprise by theoretical physicists, of their astonishment that stuff invented by mathematicians for their own delectation—often called "aesthetic"—should turn out to be such a useful tool in understanding the deep structure of the world. The most often cited text is Eugene Wigner's from 1960, the superbly titled, "The Unreasonable Effectiveness of Mathematics in the Natural Sciences." His essay moves right along to "the miracle of the appropriateness of the language of mathematics for the formulation of the laws of physics is a wonderful gift which we neither understand nor deserve." Yes, it is called a *miracle*.

Mark Steiner (1998, 13–14) begins with an anthology of similar cries of wonder from Heinrich Hertz, Stephen Weinberg (*spooky*), Richard Feynman (*amazing*), Kepler, and Roger Penrose. This is what in my artificial ordering I call an *App 2*.

DIFFERENT KINDS OF APPLICATION

Each of my six *Apps* feeds into or generates a philosophical issue about mathematics and so contributes to the perennial and central character of the philosophy of mathematics within the Western philosophical tradition. But since some of these types of application are relatively new, they have fed philosophizing only recently.

App 0: Math Applied to Math

Philosophers who discuss application seldom discuss the application of one branch of mathematics to mathematics. Yet ever since Descartes created analytic geometry by applying arithmetic to geometry, that has been one of the main engines of mathematical progress. Anyone who has read popular expositions of Andrew Wiles' proof of Fermat's last theorem recalls how it joins together previously unrelated fields of mathematics. This joining is normal in seriously novel mathematics, rather than unusual. Why should there be so much ultimate connectedness behind so much apparent diversity? If we follow the cognitive scientists who think that there are distinct mental modules for arithmetical reasoning and for spatial reasoning, Descartes' *Geometry* of 1637 is all the more astonishing. This question needs a lot of philosophical work, right now.

App 1: The Pythagorean Dream

There has long been the sense that mathematics is uncovering the deep structure of the world and that the essence of the universe just is mathematical. Kepler worked out the planetary system on Pythagorean grounds. The Book of Nature, as Galileo called it, is written in the Language of Mathematics. P.A.M. Dirac, who is certainly among the five greatest theoretical physicists of the twentieth century, had a profound conception of this sort, expressed in many places, e.g., "There is thus a possibility that the ancient dream of philosophers to connect all Nature with the properties of whole numbers will some day be realized" (Dirac 1939, 129).

In the "Pythagorean" ideal, mathematics of a deep and simple sort really just is the structure of reality, and perhaps *must* be. Physics and metaphysics merge, as they always have since Plato proposed in the *Timaeus* that the elements were made up of the five regular polyhedra. Should I not file this as Ancient, and not Enlightenment? I do not, because it was a philosophical speculation in ancient times, but only after Galileo did it seem to become a royal road to understanding the universe.

The question of why Pythagorean reasoning about the real world is so effective has been around since Galileo. Mark Steiner is the one philosopher working today who has taken it seriously. But he is not a fabulist Pythagorean. Steiner holds that the very concept of mathematics itself is "anthropocentric" (1998, 6).

App 2: Mathematical Physics

We pass to a less metaphysical way of thinking. Elegant mathematical structures are found to provide amazingly accurate models for processes found in nature. That's what led the physicist to exclaim "Spooky!" A version of this attitude is exemplified in the paper "Polyhedra in Physics, Chemistry and Geometry" by Michael Atiyah, another Fields Medalist. Polyhedra, yes, but not quite Timaean! More generally, in the survey article, "Geometry and Physics," published in 2010, Atiyah stands at the door of Wigner's "unreasonable effectiveness." Chronologically, this astonishment at the success of mathematical physics is an early twentieth century conception, while Pythagorean *App 1* is at least as old as the *Timaeus*. If you are not seduced by Pythagoras, why is mathematics so unreasonably effective?

App 3: Mission-Oriented Applied Math

In *Apps 1* and *2*, the mathematics is (as Wigner put it) often developed for its "aesthetic" interest, or at any rate with only a loose and flexible relation to material structures. Most of what we actually call applied mathematics is deliberately done for a relatively practical purpose. Let us take SIAM—the Society for Industrial and Applied Mathematics—as the prototype for *App 3*. Here is its self-definition (SIAM 2010):

> Applied mathematics is the branch of mathematics that is concerned with developing mathematical methods and applying them to science, engineering, industry, and society. It includes mathematical topics such as partial and ordinary differential equations, linear algebra, numerical analysis, operations research, discrete mathematics, optimization, control, and probability. Applied mathematics uses math-modelling techniques to solve real-world problems.

In sheer population, *App 3* totally dwarfs *App 2*. SIAM has "over 13,000 individual members. Almost 500 academic, manufacturing, research and development, service and consulting organizations, government, and military organizations worldwide are institutional members." And it mostly does "hard" or "dry" applications. There is a lot more applied mathematics in the life sciences, and that field is growing incredibly fast. When bureaucrats organize departments with the words "applied mathematics" in their titles, it is *App 3* that they have in mind. (Cambridge University does it this way: There is a Department of Pure Mathematics and Mathematical Statistics (!) and a Department of Applied Mathematics (*App 3*) and Theoretical Physics (*App 2*).)

App 4: Common or Garden

Then there are endless "common or garden" uses of mathematics by accountants, shopkeepers, carpenters, contractors, farmers, and lawyers—and by almost everybody in the contemporary world. Minimal numeracy has long been required of almost everyone who wants to have the minimum standard of living of an industrial state. When Kant (and Russell) asked how pure mathematics is possible, they were mostly thinking of *App 4*. But notice that one may be less inclined to focus on *truths* here than on *rules* for making informative transformations. Kant took for granted that "$5 + 7 = 12$" is a truth which we know to be true. His conundrum would seem less pressing if we took this formula to be an instruction about the manipulation of information. And if we think common or garden mathematics is about rules, rather than propositions, we shall be less tempted to engage in what Wittgenstein called "mathematical alchemy," which transmutes numbers into objects.

App 5: Unintended Social Uses

All of the preceding are intended uses of mathematics, even when a branch of mathematics has been developed for one purpose and turns out to be useful for another for which it was not immediately intended. There are also *unintended uses*, where purists might say the use is almost a perversion of intended uses.

Our educational systems force all children to acquire minimum mathematical skills. Those who do better, up to high school level, are advanced in the social hierarchy. It reminds us of the system of the British ruling classes, which determined who would run the empire by success in mastering Latin and Greek. One of the uses of such institutional practices is to preserve the present social order. The most famous perversion of mathematics for elitist purposes was Plato's. Every senior civil servant in his republic had to spend the best 10 years of his life mastering mathematics, not to be a better finance minister or general but to acquire a moral status.

App 6: Off the Wall

Finally, we can imagine uses of mathematics that are not just unintended by the mathematicians who do the work, but, we might say, "off the wall." One of these is Wittgenstein's; it happens to be literally on the wall (Wittgenstein 1979, 34).

> Why should not the only application of the integral and differential calculus etc. not be for patterns on wallpaper? Suppose they were invented just because people like a pattern of this kind? This would be a perfectly good application.

It is possible that the wallpaper example was intended to soften his audience up for a much less bizarre application about which he had much more to say, namely, predicting what a person (or most people) will do, when asked, say, to multiply 31 by 12. And that leads on to his reflections on following a rule. Steiner is completing a book on Wittgenstein on mathematics that focuses on just this notion of application.

It is seldom noticed that *Anwendung*—application—is, after "proof," about the most frequently used noun of content in the first edition of Wittgenstein's *Remarks on the Foundations of Mathematics (1978)*. Georg Kreisel (1958) did notice this fact in his review of the first edition and takes it to be one of the sources of what went wrong with that body of work. Steiner and I have the opposite view.

On another occasion, I argue that Wittgenstein's discussion of application, at *Apps* 4 and 6, leads into an important reflection on an aspect of Kant's *shout* that I have not had time to discuss here: the sense of compulsion, of necessity, of "the hardness of the logical *must*." Even without such an addendum, it has become clear that application, at all the *Apps* listed, is deep in the explanation of why there is philosophy of mathematics at all.

Notes

1. I use the traditional Kemp Smith translation rather than that of Guyer and Wood (Cambridge 1998) because for content it does not matter here, while for euphony I prefer "be he Thales or some other" to "whether he be called 'Thales' or had some other name." And I confess I prefer "the happy thought" to "the happy inspiration."

2. Their discussion was published in French in 1989, and a few years later in English translation, with an additional discussion. Connes, the mathematician (Fields Medal 1982), was awarded the Clay prize in 2000 "for revolutionizing the field of operator algebras, for inventing non-commutative geometry, and for discovering that these ideas appear everywhere, including the foundations of theoretical physics." Changeux is an eminent neurobiologist. The two men are colleagues at the Collège de France in Paris.

3. The English translation says "fifteen years ago," but I have corrected this; Connes wrote, "il y a une quinzaine d'années." And that was in 1989, hence my "about 1974." The time 1973–1975 was a period of intense activity whose temporary epicenter was Cambridge, England, but with much input from, for example, Michigan and Bielefeld. For a popular exposition, see Ronan 2006. Contrary to what Connes implied in 1989, the classification of finite simple groups was not proved to everyone's satisfaction until in 2004 Aschbacher and Smith published a proof in two volumes which no one person will ever, in the future, read.

4. Quoted by permission from a conversation July 27, 2010.

5. There is a translation problem here; the word is "archaïque." The English "archaic" is almost a false friend in this context. The translator of Connes and Changeux (1995) uses "archaic," while the translator of Connes, Lichnerowicz and Schützenberger (2000a) uses "primordial." There is a brief popular exposition of the idea in Connes (2000b).

6. "Proven" translates "bewährt." Previous translators (e.g., Lucas, Manchester University Press, 1953) have offered "proved." I would like to argue that Kant was well aware of the centrality of proof, but here he might have meant by "bewährt" no more than that mathematics has stood up to the tests of time. On this occasion, I have kept the traditional "knowledge" as the translation of "Erkenntis."

References

Atiyah, Michael, and Sutcliffe, Paul. 2003. "Polyhedra in physics, chemistry and geometry." *Milan J. Math* 71, 33–58.
Atiyah, Michael, Robert Dijkgraaf, and Nigel Hitchin. 2010. "Geometry and physics." *Philosophical Transactions of the Royal Society A* 368, 913–926.

Balaguer, Mark. 1998. *Platonism and Anti-Platonism in Mathematics.* New York: Oxford University Press.

Benacerraf, Paul. 1973. "Mathematical truth." *Journal of Philosophy* 70, 661–680.

Brown, Gary I. 1991. "The evolution of the term 'Mixed Mathematics.'" *Journal of the History of Ideas* 52, 81–102.

Burnyeat, Miles. 2000. "Plato on why mathematics is good for the soul." In T. J. Smiley (ed.) *Mathematics and Necessity: Essays in the History of Philosophy.* Published for the British Academy by Oxford, 2000, 1–82.

Butterworth, Brian. 1999. *The Mathematical Brain.* London: Macmillan.

Colyvan, Mark. 1998. "Indispensability arguments in the philosophy of mathematics." *Stanford Encyclopedia of Philosophy.* http://plato.stanford.edulentries/mathphil-indis/ (December 2010).

Connes, Alain, and Jean-Pierre Changeux. 1989. *Matière à pensée.* Paris: Odile Jacob.

Connes, Alain, and Jean-Pierre Changeux. 1995. *Conversations on Mind, Matter, and Mathematics.* Translation of 1989. Princeton, NJ: Princeton University Press.

Connes, Alain, André Lichnerowicz, and M. P. Schützenberger. 2000a. *Triangle of Thoughts.* Providence, RI: American Mathematical Society.

Connes, Alain. 2000b. "La réalité mathématique archaïque." *La Recherche,* No. 332, 109.

Dehaene, Stanislas. 1997. *The Number Sense: How the Mind Creates Mathematics.* New York: Oxford University Press.

Dehaene, Stanislas. 2009. "Origins of mathematical intuitions: The case of arithmetic." *Annals of the New York Academy of Science,* 1156, 232–259.

Dirac, P.A.M. 1939. "The relation between mathematics and physics." *Proceedings of the Royal Society (Edinburgh)* 59, II, 122–129. (Reprinted *Collect Works,* Cambridge, 1995, 907–914.)

Dummett, Michael. 1991. *Frege's Philosophy of Mathematics.* Cambridge, MA: Harvard University Press.

Gowers, Timothy. 2002. *Mathematics: A Very Short Introduction.* New York: Oxford University Press.

Gowers, Timothy. 2006. "Does mathematics need a philosophy?" In Reuben Hersch (ed.) *18 Unconventional Essays on the Nature of Mathematics.* New York: Springer Science+Business Media, 182–201.

Hardy, G. H. 1929. "Mathematical proof." *Mind* 37, 1–25.

Hempel, C. G. 1949. "On the nature of mathematical truth." In H. Feigl and W. Sellars (eds.), *Readings in Philosophical Analysis.* New York: Appleton Century Crofts, 1949, 222–237. First published in *American Mathematical Monthly* 52 (1949).

Kant, Immanuel. 1930. *Critique of Pure Reason.* (Norman Kemp Smith, trans.) London: Macmillan.

Kant, Immanuel. 1997. *Prolegomena to Any Future Metaphysics That Will Be Able to Come Forward as Science.* (Gary Hatfield, trans.) Cambridge, MA: Cambridge University Press.

Kreisel, Georg. 1958. "Wittgenstein's *Remarks on the Foundations of Mathematics.*" *British Journal for the Philosophy of Science,* 9, 135–158.

Lloyd, Geoffrey. 1990. *Demystifying Mentalities,* Cambridge, MA: Cambridge University Press.

Maddy, Penelope. 2008. "How applied mathematics became pure." *The Review of Symbolic Logic* 1, 16–41.

Quine, W.V.O. 1950. *Methods of Logic.* New York: Holt.

Quine, W.V.O. 2008. "Naturalism; Or, living within one's means." In D. Føllesdal et al. (eds.), *Confessions of a Confirmed Extensionalist and other Essays,* Cambridge, MA: Harvard University Press, 461–472. (Reprinted from *Dialectica* 49 (1995), 251–261.)

Netz, Reviel. 1999. *The Shaping of Deduction in Greek Mathematics: A Study in Cognitive History.* Cambridge, UK: Cambridge University Press.

Ronan, Mark. 2006. *Symmetry and the Monster: One of the Greatest Quests in Mathematics*. New York: Oxford University Press.

Russell, Bertrand. 1912. *The Problems of Philosophy*. London: Macmillan.

SIAM. 2010. <http://www.siam.org/about/more/whatis.php> (Dec. 27, 2010).

Spelke, Elizabeth, S. A. Lee, and V. Izard. 2010. "Beyond core knowledge: Natural geometry." *Cognitive Science*, 34(5), 863–884.

Steiner, Mark. 1998. *The Applicability of Mathematics as a Philosophical Problem*. Cambridge, MA: Harvard University Press.

Wigner, Eugene. 1960. "The unreasonable effectiveness of mathematics in the natural sciences." *Communications in Pure and Applied Mathematics*, 13.

Wittgenstein, Ludwig. 1978. *Remarks on the Foundations of Mathematics*. Edited by G.E.M. Anscombe. Oxford, U.K.: Blackwell.

Wittgenstein, Ludwig. 1979. *Wittgenstein's Lectures, Cambridge 1932–1935*. Edited by Alice Ambrose. Totawa, NJ: Rowman and Littlefield.

Ultimate Logic: To Infinity and Beyond

RICHARD ELWES

When David Hilbert left the podium at the Sorbonne in Paris, France, on August 8, 1900, few of the assembled delegates seemed overly impressed. According to one contemporary report, the discussion following his address to the second International Congress of Mathematicians was "rather desultory." Passions seem to have been more inflamed by a subsequent debate on whether Esperanto should be adopted as mathematics' working language.

Yet Hilbert's address set the mathematical agenda for the twentieth century. It crystallized into a list of 23 crucial unanswered questions, including how to pack spheres to make best use of the available space, and whether the Riemann hypothesis, which concerns how the prime numbers are distributed, is true.

Today many of these problems have been resolved, sphere-packing among them. Others, such as the Riemann hypothesis, have seen little or no progress. But the first item on Hilbert's list stands out for the sheer oddness of the answer supplied by generations of mathematicians since: that mathematics is simply not equipped to provide an answer.

This curiously intractable riddle is known as the continuum hypothesis, and it concerns that most enigmatic quantity, infinity. Now, 140 years after the problem was formulated, a respected U.S. mathematician believes he has cracked it. What's more, he claims to have arrived at the solution not by using mathematics as we know it, but by building a new, radically stronger logical structure: a structure he dubs "ultimate L."

The journey to this point began in the early 1870s, when the German Georg Cantor was laying the foundations of set theory. Set theory deals with the counting and manipulation of collections of objects and provides the crucial logical underpinnings of mathematics: Because numbers can be associated with the size of sets, the rules for manipulating sets also determine the logic of arithmetic and everything that builds on it.

These dry, slightly insipid, logical considerations gained a new tang when Cantor asked a critical question: How big can sets get? The obvious answer—infinitely big—turned out to have a shocking twist: Infinity is not one entity, but comes in many levels.

How so? You can get a flavor of why by counting up the set of whole numbers: 1, 2, 3, 4, 5 . . . How far can you go? Why, infinitely far, of course—there is no biggest whole number. This is one sort of infinity, the smallest, "countable" level, where the action of arithmetic takes place.

Now consider the question "How many points are there on a line?" A line is perfectly straight and smooth, with no holes or gaps; it contains infinitely many points. But this is not the countable infinity of the whole numbers, where you bound upward in a series of defined, well-separated steps. This is a smooth, continuous infinity that describes geometrical objects. It is characterized not by the whole numbers but by the real numbers: the whole numbers plus all the numbers in between that have as many decimal places as you please: $0.1, 0.01, \sqrt{2}, \pi$, and so on.

Cantor showed that this "continuum" infinity is in fact infinitely bigger than the countable, whole-number variety. What's more, it is merely a step in a staircase leading to ever-higher levels of infinities stretching up as far as, well, infinity.

Although the precise structure of these higher infinities remained nebulous, a more immediate question frustrated Cantor. Was there an intermediate level between the countable infinity and the continuum? He suspected not but was unable to prove it. His hunch about the non-existence of this mathematical mezzanine became known as the continuum hypothesis.

Attempts to prove or disprove the continuum hypothesis depend on analyzing all possible infinite subsets of the real numbers. If every subset is either countable or has the same size as the full continuum, then it is correct. Conversely, even one subset of intermediate size would render it false.

A similar technique using subsets of the whole numbers shows that there is no level of infinity below the countable. Tempting as it might be to think that there are half as many even numbers as there are whole numbers in total, the two collections can in fact be paired off exactly. Indeed, every set of whole numbers is either finite or countably infinite.

Applied to the real numbers, though, this approach bore little fruit, for reasons that soon became clear. In 1885, the Swedish mathematician Gosta Mittag-Leffler had blocked publication of one of Cantor's papers on the basis that it was "about 100 years too soon" (Dauben 1990). And as the British mathematician and philosopher Bertrand Russell showed in 1901, Cantor had indeed jumped the gun. Although his conclusions about infinity were sound, the logical basis of his set theory was flawed, resting on an informal and ultimately paradoxical conception of what sets are.

It was not until 1922 that two German mathematicians, Ernst Zermelo and Abraham Fraenkel, devised a series of rules (Halmos 1960) for manipulating sets that was seemingly robust enough to support Cantor's tower of infinities and stabilize the foundations of mathematics. Unfortunately, though, these rules delivered no clear answer to the continuum hypothesis. In fact, they seemed strongly to suggest that there might even not be an answer.

Agony of Choice

The immediate stumbling block was a rule known as the "axiom of choice." It was not part of Zermelo and Fraenkel's original rules but was soon bolted on when it became clear that some essential mathematics, such as the ability to compare different sizes of infinity, would be impossible without it.

The axiom of choice states that if you have a collection of sets, you can always form a new set by choosing one object from each of them. That sounds anodyne, but it comes with a sting: You can dream up some twisted initial sets that produce even stranger sets when you choose one element from each. The Polish mathematicians Stefan Banach and Alfred Tarski soon showed how the axiom could be used to divide the set of points defining a spherical ball into six subsets, which could then be slid around to produce two balls of the same size as the original (Stromberg 1979). That was a symptom of a fundamental problem: The axiom allowed peculiarly perverse sets of real numbers to exist whose properties could never be determined. If so, this was a grim portent for ever proving the continuum hypothesis.

This news came at a time when the concept of "unprovability" was just coming into vogue. In 1931, the Austrian logician Kurt Gödel proved his notorious "incompleteness theorem" (Davis 2006). It shows that even with the most tightly knit basic rules, there are always statements about sets or numbers that mathematics can neither verify nor disprove.

At the same time, though, Gödel had a crazy-sounding hunch about how you might fill in most of these cracks in mathematics' underlying logical structure: You simply build more levels of infinity on top of it. That method goes against anything we might think of as a sound building code, yet Gödel's guess turned out to be inspired. He proved his point in 1938. By starting from a simple conception of sets compatible with Zermelo and Fraenkel's rules and then carefully tailoring its infinite superstructure, he created a mathematical environment in which both the axiom of choice and the continuum hypothesis are simultaneously true. He dubbed his new world the "constructible universe"—or simply "L."

L was an attractive environment in which to do mathematics, but there were soon reasons to doubt it was the "right" one. For a start, its infinite staircase did not extend high enough to fill in all the gaps known to exist in the underlying structure. In 1963, Paul Cohen of Stanford University in California put things into context when he developed a method for producing a multitude of mathematical universes to order, all of them compatible with Zermelo and Fraenkel's rules.

This method was the beginning of a construction boom. "Over the past half-century, set theorists have discovered a vast diversity of models of set theory, a chaotic jumble of set-theoretic possibilities," says Joel Hamkins at the City University of New York. Some are "L-type worlds" with superstructures like Gödel's L, differing only in the range of extra levels of infinity they contain; others have wildly varying architectural styles with completely different levels and infinite staircases leading in all sorts of directions.

For most purposes, life within these structures is the same: Most everyday mathematics does not differ between them, nor do the laws of physics. But the existence of this mathematical "multiverse" also seemed to dash any notion of ever getting to grips with the continuum hypothesis. As Cohen was able to show ([1966] 2008), in some logically possible worlds the hypothesis is true and there is no intermediate level of infinity between the countable and the continuum; in others, there is one; in still others, there are infinitely many. With mathematical logic as we know it, there is simply no way of finding out which sort of world we occupy.

That's where Hugh Woodin of the University of California, Berkeley, has a suggestion (2010). The answer, he says, can be found by stepping outside our conventional mathematical world and moving on to a higher plane.

Woodin is no "turn on, tune in" guru. A highly respected set theorist, he has already achieved his subject's ultimate accolade: a level on the infinite staircase named after him. This level, which lies far higher than anything envisaged in Gödel's L, is inhabited by gigantic entities known as Woodin cardinals.

Woodin cardinals illustrate how adding penthouse suites to the structure of mathematics can solve problems on less rarefied levels below. In 1988, the U.S. mathematicians Donald Martin and John Steel (1989) showed that if Woodin cardinals exist, then all "projective" subsets of the real numbers have a measurable size. Almost all ordinary geometrical objects can be described in terms of this particular type of set, so this proof was just the buttress needed to keep uncomfortable apparitions such as Banach and Tarski's ball out of mainstream mathematics.

Such successes left Woodin unsatisfied, however. "What sense is there in a conception of the universe of sets in which very large sets exist, if you can't even figure out basic properties of small sets?" he asks. Even 90 years after Zermelo and Fraenkel had supposedly fixed the foundations of mathematics, cracks were rife. "Set theory is riddled with unsolvability. Almost any question you want to ask is unsolvable," says Woodin. And right at the heart of that statement lay the continuum hypothesis.

Ultimate L

Woodin and others spotted the germ of a new, more radical approach while investigating particular patterns of real numbers that pop up in various L-type worlds. The patterns, known as universally Baire sets, subtly changed the geometry possible in each of the worlds and seemed to act as a kind of identifying code for it. And the more Woodin looked, the more it became clear that relationships existed between the patterns in seemingly disparate worlds. By patching the patterns together, the boundaries that had seemed to exist between the worlds began to dissolve, and a map of a single mathematical superuniverse was slowly revealed. In tribute to Gödel's original invention, Woodin dubbed this gigantic logical structure "ultimate L."

Among other things, ultimate L provides for the first time a definitive account of the spectrum of subsets of the real numbers: For every forking point between worlds that Cohen's methods open up, only one possible route is compatible with Woodin's map. In particular, it implies Cantor's hypothesis to be true, ruling out anything between countable infinity and the continuum. That would mark not only the end of a 140-year-old conundrum but also a personal turnaround for Woodin: 10 years ago, he was arguing that the continuum hypothesis should be considered false.

Ultimate L does not rest there. Its wide, airy space allows extra steps to be bolted to the top of the infinite staircase as necessary to fill in gaps below, making good on Gödel's hunch about rooting out the unsolvability that riddles mathematics. Gödel's incompleteness theorem would not be dead, but you could chase it as far as you pleased up the staircase into the infinite attic of mathematics.

The prospect of finally removing the logical incompleteness that has bedevilled even basic areas such as number theory is enough to get many mathematicians salivating. There is just one question. Is ultimate L ultimately true?

Andrés Caicedo, a logician at Boise State University in Idaho, is cautiously optimistic. "It would be reasonable to say that this is the 'correct' way of going about completing the rules of set theory," he says (2010). "But there are still several technical issues to be clarified before saying confidently that it will succeed."

Others are less convinced. Hamkins (forthcoming), who is a former student of Woodin's, holds to the idea that there simply are as many legitimate logical constructions for mathematics as we have found so far. He thinks mathematicians should learn to embrace the diversity of the mathematical multiverse, with spaces where the continuum hypothesis is true and others where it is false. The choice of which space to work in would then be a matter of personal taste and convenience. "The answer consists of our detailed understanding of how the continuum hypothesis both holds and fails throughout the multiverse," he says.

Woodin's ideas need not finish off this choice entirely, though: Aspects of many of these diverse universes will survive inside ultimate L. "One goal is to show that any universe attainable by means we can currently foresee can be obtained from the theory," says Caicedo (2010). "If so, then ultimate L is all we need."

In 2010, Woodin presented his ideas to the same forum that Hilbert had addressed more than a century earlier, the International Congress of Mathematicians, this time in Hyderabad, India. Hilbert famously once defended set theory by proclaiming "No one shall expel us from the paradise that Cantor has created." But we have been stumbling around that paradise with no clear idea of where we are. Perhaps now a guide is within our grasp—one that will take us through this century and beyond.

References

Caicedo, Andrés. 2010. Mathoverflow website <http://mathoverflow.net/questions/ 46907/ completion-of-zfc/46920#46920> and e-mail correspondence (homepage, http://math .boisestate.edu/~caicedo/).

Cohen, Paul J. (1966) 2008. *Set Theory and the Continuum Hypothesis*. Mineola, NY: Dover Books on Mathematics.

Dauben, Joseph. 1990. *Georg Cantor*. Princeton, NJ: Princeton University Press. This book may be a useful source.

Davis, Martin. 2006. "The Incompleteness Theorem." *Notices of the AMS* 53 (4): 414. This article may be useful for further reading.

Halmos, Paul R. 1960. *Naive Set Theory*, Princeton, NJ: Van Nostrand. This is an introductory book to this branch of mathematics.

Hamkins, Joel D. Forthcoming. "The Set-Theoretical Multiverse." *Review of Symbolic Logic*.

Martin, Donald, and John Steel. 1989. "A Proof of Projective Determinacy." *Journal of the AMS 2 (1)*.

Stromberg, Karl. 1979. "The Banach–Tarski Paradox." *The American Mathematical Monthly* 86 (3): 151–161. This article may be useful for further reading.

Woodin, Hugh. 2010. "Strong Axioms of Infinity and the Search for V." *Proceedings of the International Congress of Mathematicians*, Hyderabad, India, World Scientific Publishing Company.

Mating, Dating, and Mathematics: It's All in the Game

MARK COLYVAN

Why do people stay together in monogamous relationships? Love? Fear? Habit? Ethics? Integrity? Desperation? In this chapter, I consider a rather surprising answer that comes from mathematics. It turns out that cooperative behavior, such as mutually faithful marriages, can be given a firm basis in a mathematical theory known as *game theory*. I suggest that faithfulness in relationships is fully accounted for by narrow self-interest in the appropriate game theory setting. This is a surprising answer because faithful behavior is usually thought to involve love, ethics, and caring about the well-being of your partner. It seems that the game theory account of faithfulness has no need for such romantic notions. I consider the philosophical upshot of the game theoretic answer and see if it really does deliver what is required. Does the game theoretic answer miss what is important about faithful relationships or does it help us get to the heart of the matter? Before we start looking at lasting, faithful relationships, though, let's get a feel for how mathematics might be used to help in matters of the heart. Let's first consider how mathematics might shed light on dating to find a suitable partner.

A Lover's Question

Consider the question of how many people you should date before you commit to a more permanent relationship, such as marriage. Marrying the first person you date is, as a general strategy, a bad idea. After all, there's very likely to be someone better out there, but by marrying too early you're cutting off such opportunities. But at the other extreme, always leaving your options open by endlessly dating and continually looking for someone better is not a good strategy either. It would seem that somewhere between marrying your first high school crush and dating forever lies the ideal strategy. Finding this ideal strategy is an optimization problem and, believe it or not, is particularly amenable to mathematical treatment. In fact, if we add a couple of constraints to the problem, we have the classic mathematical problem known as the *secretary problem.*

The mathematical version of the problem is presented as one of finding the best secretary (which is just a thin disguise for finding the best mate) by interviewing (i.e., dating) a number of applicants. In the standard formulation, you have a finite and known number of applicants, and you must interview these n candidates sequentially. Most importantly, you must decide whether to accept or reject each applicant immediately after interviewing him or her; you cannot call back a previously interviewed applicant. This method makes little sense in the job search context but is natural in the dating context: Typically, boyfriends and girlfriends do not take kindly to being passed over for someone else and are not usually open to the possibility of a recall. The question, then, is how many of the n possible candidates should you interview before making an appointment? Or in the dating version of the problem, the question is how many people should you date before you marry?

It can be shown mathematically that the optimal strategy, for a large applicant pool (i.e., when n is large) is to pass over the first n/e (where e is the transcendental number from elementary calculus—the base of the natural logarithm, approximately 2.718) applicants and accept the next applicant who's better than all those previously seen. This method gives a probability of finding the best secretary (mate) at n/e, or approximately 0.37. For example, suppose that there are 100 eligible partners in your village, tribe, or social network; this strategy advises you to sample the population by dating the first 37, then choose the first after that who's better than all who came before. Of course, you might be unlucky in a number of ways. For example, the perfect mate might be in the first 37 and get passed over during the sampling phase. In this case, you continue dating the rest but find no one suitable and grow old alone, dreaming of what might have been. Another way you might be unlucky is if you have a run of really weak candidates in the first 37. If the next few are also weak but there's one who's better than the first 37, you commit to that one and find yourself in a suboptimal marriage. But the mathematics shows that even though things can go wrong in these ways, the strategy outlined here is still the best you can do. The news gets worse, though: Even if you stringently follow this best strategy, you still only have a bit better than a one in three chance of finding your best mate.[1]

This problem and its mathematical treatment are instructive in a number of ways. Here I want to draw attention to the various idealizations and assumptions of this way of setting things up. Notice that we started with a more general problem of how many people you should date before you marry, but in the mathematical treatment, we stipulate that the population of eligible partners is fixed and known. It's interesting that the size of this population does not change the strategy or your chances of finding your perfect partner—the strategy is as I just described and, so long as the population is large, the probability of success remains at 0.37. The size of the population just affects the number of people in the initial sample. Still, stipulating that the population is fixed is an idealization. Most pools of eligible partners are not fixed in this way—we meet new people, and others who were previously in relationships later become available, and others who were previously available enter new relationships and become unavailable. In reality, the population of eligible candidates is not fixed, but is open ended and in flux.

The mathematical treatment also assumes that the aim is to marry the best candidate. This, in turn, has two further assumptions. First, it assumes that it is in fact possible to rank candidates in the required way and that you will be able to arrive at this ranking via one date with each. We can have ties between candidates, but we are not permitted to have cases where we cannot compare candidates. The mathematical treatment also assumes that we're after the *best* candidate and anything less than this is a failure. For instance, if you have more modest goals and are only interested in finding someone who'll meet a minimum standard, you need to set things up in a completely different way—it then becomes a satisficing problem and is approached quite differently.

Another idealization of the mathematical treatment—and this is the one I am most interested in—is that finding a partner is assumed to be one-sided. The treatment we're considering here assumes that it is an employers' market. It assumes, in effect, that when you decide that you want to date someone, he or she will agree, and that when you decide to enter a relationship with someone, again, the person will agree. This mathematical equivalent of wishful thinking makes the problem more tractable but is, as we all know, unrealistic.

A natural way to get around this last idealization is to stop thinking about your candidate pool as a row of wallflowers at a debutants' ball, and instead think of your potential partners as active agents engaged in their own search for the perfect partner. The problem, thus construed, becomes much more dynamic and much more interesting. It becomes one of coordinating strategies. There is no use setting your sights on a partner who will not reciprocate. In order for everyone to find someone to reciprocate their interest, a certain amount of coordination between parties is required. This brings us to game theory.

The Game of Love

Game theory is the study of decisions where one person's decision depends on the decisions of other people.[2] Think of games like chess or tennis, where your move is determined, at least in part, by what you think the other player's response will be. It is important to note that games do not have to be fun and are not, in general, mere diversions. The Cold War arms race can be construed as a "game" (in this technical sense of game) between military powers, each second-guessing what the other would do in response to their "moves." Indeed, the Cold War was the stage for one of the original and most important applications of game theory. The basic idea of game theory is quite simple and should be familiar: A number of players are making decisions, each of which depends on the decisions of the other players.

It's probably best to illustrate game theory via an example. Let's start with the *stag hunt*. This game originates in a story of cooperative hunting by the eighteenth-century political philosopher Jean-Jacques Rousseau.[3] In its simplest form, the game consists of two people setting out to hunt a stag. It will take the cooperation of both to succeed in the hunt, and the payoff for a successful stag hunt is a feast for all. But each hunter is tempted by lesser prey: a hare, for example. If one of the hunters defects from the stag hunt and opportunistically hunts a passing hare, the defector is rewarded, but the stag hunt fails so that the nondefector is not rewarded. In decreasing order of preference, the rewards are: stag, hare, and nothing. So the cooperative outcome (both hunt stag) has the maximum payoff for each of the hunters, but it is unstable in light of the ever-present temptation for each hunter to defect and hunt hare instead. Indeed, hunting hare is the safer option. In the jargon of game theory, the cooperative solution of hunting stag is *Pareto optimal* (i.e., there is no outcome that is better for both hunters), while the mutual defect solution is *risk dominant* (in that it does not leave you empty-handed if your fellow hunter decides to defect and hunt hare), but it is not Pareto optimal. That is, the cooperative solution is best for both hunters and given that the other party cooperates in the stag hunt, then you should too. But if the other party defects and hunts hare, then so should you. Most importantly, both these outcomes are stable, since neither party will unilaterally change from cooperation to defection or from defection to cooperation (again, in the jargon of game theory, the mutual defect and cooperation solutions are *Nash equilibria*).[4] So, in particular, if you both play it safe and hunt hares, there seems no easy way to get to the mutually preferable cooperative solution of stag hunting. Cooperation seems both hard to achieve and somewhat fragile. This game is important because it is a good model of many forms of cooperative behavior.[5]

Consider another example, just to get a feel for game theory: the *prisoner's dilemma*. The scenario here is one where two suspects are questioned separately by the police and each suspect is invited to confess to a crime the two have jointly committed. There is not sufficient evidence for a conviction, so each suspect is offered the following deal: If one confesses, that suspect will go free while the other serves the maximum sentence; if they both confess, they will both serve something less than the maximum sentence; if neither confesses, they will both be charged with minor offenses and receive sentences less than any of those previously mentioned. In order of preference, then, each suspect would prefer (1) to confess while the other does not confess, (2) that neither confess, (3) that both confess, and (4) not to confess while the other confesses. Put like this, it is clear what you should do: You should confess to the crime. Why? Because, irrespective of what the other suspect does, you will be better off if you confess. But here's the problem: If both suspects think this way, as surely they should, they will both end up with the second worst outcome (3). As a pair, their best outcome is (2)—this is Pareto optimal, since neither can do better than this without the other doing worse—but the stable solution is (3) where both defect—this is the Nash equilibrium, since given that one confesses the other should too. Group rationality and individual rationality seem to come apart. Individual rationality recommends both confessing, even though this is worse for both parties than neither confessing. Again, we see that defection (this time from any prearranged agreement between the suspects to not confess) is rewarded and cooperation is fragile.[6]

What has hunting stags and police interrogations got to do with dating—crude metaphors aside? First, these two games demonstrate how important it is to consider the decisions of others when making your own decisions. What you do is determined, in part at least, by what the other players in the game do and vice versa. So, too, with relationships. In fact, the stag hunt is a good model of cooperation in a relationship. Think of cooperatively hunting stag as staying faithful in a monogamous relationship. When all is going well, this cooperation holds great benefits for both parties. But there is always the temptation for one partner to opportunistically defect from the relationship, to have an affair. This is the "hunting hare" option. If both partners do this, we have mutual defection, where both parties defect from the relationship in favor of affairs. This game theoretic way of looking at things gives us a very useful framework for thinking about our original question of why people stay in monogamous relationships.

Where Did Our Love Go?

We are now in a position to see one account of how monogamous relationships are able to persist. Sometimes it is simply the lack of opportunity for outside affairs. After all, there's no problem seeing why people cooperate in hunting stags when there are no alternatives. The more interesting case is when there *are* other opportunities. According to the game theory account we are interested in here, an ongoing monogamous relationship is a kind of social contract and is akin to the agreement to mutually hunt stag. But what binds one to abiding by this contract when there are short-term unilateral gains for defecting? Indeed, it seems that game theory suggests defection as a reasonable course of action in such situations. If the chances of catching a stag (or seeing the benefits of a lasting monogamous relationship) are slim, defecting by opportunistically catching a hare (or having an affair) seems hard to avoid, perhaps even prudent. But we must remember that the games in question are not isolated one-off situations, and this fact is key.

Although defection in the prisoner's dilemma or the stag hunt may be a reasonable course of action if the situation in question is not repeated, in cases where the game is played on a regular basis, there are much better long-term strategies. For instance, both players should see the folly of defecting in the first game, if they know that they will be repeatedly playing the same player. A better strategy is to cooperate at first and retaliate with a defection if the other player defects. Such so-called tit-for-tat strategies do well in achieving cooperation. If both players are known to be playing this strategy, they are more inclined to cooperate indefinitely. There are other good strategies that encourage cooperation in these repeated games, but the tit-for-tat strategy illustrates the point. In short, cooperation is easier to secure when the games in question are repeated, and the reason is quite simple: The long-term rewards are maximized by cooperating, even though there is the temptation of a short-term reward for defection. It's the prospect of future games that ensures cooperation now. Robert Axelrod calls this "the shadow of the future"[7] hanging over the decision. This shadow changes the relevant rewards in a way that ensures cooperation.

We can make the cooperative outcome even more likely and more stable by sending out signals about our intentions to retaliate if we ever encounter a defector. In the stag hunt, we might make it clear that defection by the other party will result in never cooperating with them again in a stag hunt. (Translated into the monogamous relationship version, this amounts to divorce or sleeping on the couch for the rest of your life.) We might even make such agreements binding by making the social contract in question public and inviting public scorn on defectors. All this threatening amounts to a change in the payoffs for the game so that defection carries with it some serious costs, costs not present in the simple one-off presentation with which we started.

It is interesting to notice that this is pretty much what goes on in the relationship case. We have public weddings to announce to our friends and the world the new social contract in place (thus increasing the cost of a possible defection); we, as a society, frown on extramarital affairs (unless they are by mutual consent); and most important of all, we are aware of the long-term payoffs of a good, secure, long-term, monogamous relationship (if, indeed, that is what is wanted).

Now it seems that we have the makings of an explanation of such relationships in terms of self-interest. Although cooperation might look as though it has to do with love, respect, ethics, loyalty, integrity, and the like, the game theory story is that it's all just narrow self-interest. It's not narrow in the sense of being shortsighted but in the sense that there's no need to consider the interests of others, except insofar as they affect oneself. As David Hume puts it, "I learn to do service to another, without bearing him any real kindness; because I foresee that he will return my service, in expectation of another of the same kind."[8] In particular, there seems to be no place for love (and acting out of love) in the account outlined here.

Love Is Strange

If all I've said so far is right, it looks as though we can explain faithful relationships in terms of narrow self-interest. It's a case of "this is good for me; who cares about you?" According to the game theory story, a faithful relationship is just a particular form of social cooperation. And all that is needed to keep the cooperation in place is mutual self-interest. It has nothing to do with right or wrong, or caring for your partner. It's all in the game and the focus on payoffs to the individual— or at least, payoffs to the individual plus the shadow of the future. We might still frown upon noncooperation but not for the reasons usually assumed. We, as a society, frown on defectors because that's also part of the game, and it's an important part of what is required to keep cooperation alive in the society at large.

You might be skeptical of all this. You might think that people fall in love and enter a relationship, not because they can get something out of it but... well, why? If you're not getting something out of it, surely you're doing it wrong! Okay, perhaps you get something out of it, but you stay committed through the hard times, through the arguments, through your partner's bad moods, not *purely* because it's good for you. You stick with a partner because he or she needs you and you're a good person, right? It might help if that's what you believe, but one take-home message from the account I'm offering here is that there's no need for anything outside the game. We don't need to entertain

anything other than self-interest as a motivation for monogamous relationships. It may well be that it's useful to believe in such things as
loyalty, goodness, and perhaps even altruism, but all that might be just
useful fictions—a kind of make-believe that's important, perhaps even
indispensable, but make-believe all the same.

Let's look at these issues in terms of ethics. The game theory account not only leaves no room for love and romance, it also seems to
leave ethics out of the picture. You might think that staying faithful is
ethically right and engaging in extramarital affairs is *unethical*. Insofar as
game theory says nothing about ethics, it would seem that it cannot be
the whole story. But we can take this same game theoretic approach
to ethics. Ethics can be thought of as a series of cooperation problems.
Thus construed, ethics is arguably explicable in the same terms.[9] The
idea is that ethical behavior is just stable, mutually beneficial behavior
that is the solution to typical coordination problems (basically ethics
is just a matter of "don't hurt me and I won't hurt you"), and societ-
ies that have robust solutions to such coordination problems do better
than those that don't have such solutions. As in the relationship case,
it might be beneficial to engage in the pretense that some actions really are right and some really are wrong, but again such pretense will
be just a further part of the game. This new twist about ethics either
makes your concerns about the dating and relationships case a lot worse
or a lot better, depending on your point of view. On the one hand, this
broader game theoretic story about ethics allows that there is room for
ethics in dating and relationships. But the ethics in question is just more
game theory.

The picture of relationships I'm sketching here might seem rather
different from the one we find in old love songs and elsewhere. I think
the difference, though, is more one of emphasis. Think of the picture
offered here as a new take on those old love songs rather than a different kind of song altogether. All the usual ingredients are here, but in an
unfamiliar form. We have fidelity, but it's there as a vehicle for serving
self-interest; ethical considerations are also there, but they too are not
what they first seem. I suggested that all the romance and ethics might
be merely a kind of make-believe, but perhaps that's overstating the
case. The pretense may run very deep, and it plausibly has a biologi-
cal basis. If this notion is right, the game theory picture can be seen as
offering insight into the true nature of romantic relationships. Love,

for instance, is seen as a commitment to cooperate on a personal level with someone, and it licenses socially acceptable forms of retribution if defection occurs. Perhaps this conception of love doesn't sound terribly romantic and is unlikely to find its way into love songs, but to my ears, this is precisely what all the songs are about—you just need to listen to them the right way. Love is less about the meeting of souls and more about the coordination of mating strategies. If this makes love sound strange, then so be it: Love is strange.

Notes

1. For more on the secretary problem, see Thomas S. Ferguson, "Who Solved the Secretary Problem?" *Statistical Science* 4 (1989), 282–296; for the many fascinating connections between mathematics and relationships, see Clio Cresswell, *Mathematics and Sex* (Sydney: Allen and Unwin, 2003).

2. For classic treatments of game theory, see John von Neumann and Oskar Morgenstern, *Theory of Games and Economic Behavior*, 2nd ed. (Princeton, NJ: Princeton University Press, 1947); and R. Duncan Luce and Howard Raiffa, *Games and Decisions: Introduction and Critical Survey* (New York: John Wiley, 1957).

3. Jean-Jacques Rousseau, *A Discourse on Inequality*, trans. M. Cranston (New York: Penguin, 1984).

4. Named after John Nash, the subject of the Sylvia Nasar book *A Beautiful Mind* (New York: Simon and Schuster, 1998) and the Ron Howard movie of the same name based on the Nasar book.

5. See Brian Skyrms, *The Stag Hunt and the Evolution of the Social Contract* (Cambridge, U.K.: Cambridge University Press, 2004). This is an excellent treatment of the stag hunt and its significance for social cooperation.

6. See William Poundstone, *Prisoner's Dilemma* (New York: Doubleday, 1992). This book is an accessible introduction to the prisoner's dilemma and game theory. It outlines the origins of game theory in the RAND Corporation during the Cold War (with a frightening application to the nuclear arms race).

7. Robert Axelrod, *The Evolution of Cooperation* (New York: Basic Books, 1984).

8. David Hume, *A Treatise of Human Nature*, ed. L. A. Selby-Bigge (Oxford, U.K.: Clarendon Press, 1949), p. 521.

9. Richard Joyce, *The Evolution of Morality* (Cambridge, MA: MIT Press, 2006).

Contributors

Gerald L. Alexanderson is the Valeriote Professor of Science at Santa Clara University in California. A former secretary and, later, president of the Mathematical Association of America (MAA), he has authored or edited 16 books, including the widely read Mathematical People series of interviews and profiles of mathematicians (with Donald J. Albers), the most recent of which was published by Princeton University Press in 2011. A former editor of *Mathematics Magazine* and the MAA's Spectrum book series, he has in recent years written a series of short historical articles to accompany cover art for the *Bulletin of the American Mathematical Society*, of which the essay contained in this volume is one.

John C. Baez is a mathematical physicist who teaches at the University of California, Riverside. In his work on quantum gravity, he introduced the concept of a "spin foam." More recently, he has studied applications of category theory to physics, and now he is studying its applications to biology, chemistry, and engineering.

sarah-marie belcastro is a free-range mathematician whose primary mathematical research area is topological graph theory. She enjoys connecting people to each other, connecting ideas to each other, and connecting people to ideas. Among her many non-pure-mathematics interests are the mathematics of knitting, pharmacokinetics, dance (principally ballet and modern), and changing the world. belcastro did her undergraduate work at Haverford College and her Ph.D. at the University of Michigan. She co-edited the books *Making Mathematics with Needlework: Ten Papers and Ten Projects* and *Crafting by Concepts: Fiber Arts and Mathematics* with Carolyn Yackel. Her latest wacky project is the introductory textbook *Discrete Mathematics with Ducks.* You may find tons of information (about her, and about other things) at her website http://www.toroidalsnark.net

Giuseppe Bruno is an associate professor in operational research and decision science methodologies at the University of Naples Federico II and an adjunct professor of operational research at the Parthenope University of Naples. He is a member of the Department of Management Engineering at the University of Naples Federico II. He holds a Ph.D. in computer science and robotics from the University of Naples Federico II. His main research interests lie in locational analysis, decision support systems for complex logistics problems, transportation science, and multicriteria analysis. He is the author of many papers, and his name has appeared in international journals and books. Currently, he is a member of the board of the Italian Association of Operational Research (AIRO), of the European Working Group on Locational Analysis, and of the European Working Group on Transportation.

Mark Colyvan is a professorial research fellow in philosophy and director of the Sydney Centre for the Foundations of Science at the University of Sydney. His publications include *The Indispensability of Mathematics* (Oxford University Press, 2001), *An Introduction to the Philosophy of Mathematics* (Cambridge University Press, 2012), and, with Lev Ginzburg, *Ecological Orbits: How Planets Move and Populations Grow* (Oxford University Press, 2003).

Brent Davis is a professor and a distinguished research chair in mathematics education at the University of Calgary. His research is developed around the educational relevance of developments in the cognitive and complexity sciences, with a particular focus on mathematical understanding. Davis has published in the areas of mathematics learning and teaching, curriculum theory, teacher education, epistemology, and action research in journals that include *Science*, *Harvard Educational Review*, *Educational Theory*, and *Qualitative Studies in Education*. His most recent books include *Engaging Minds: Changing Teaching in Complex Times* (2nd edition, 2008; co-authored with Dennis Sumara and Rebecca Luce-Kapler).

Richard Elwes is a teaching fellow at the University of Leeds. His research interests are in mathematical logic and its applications. Since winning the *Plus magazine* New Writer's competition in 2006, he has devoted much of his time to disseminating mathematical ideas in print and in person. He is a regular feature writer for the *New Scientist* magazine and the author of several popular books, including *Mathematics 1001* (*Quercus*, 2010). For more information, see www.richardelwes.co.uk

Susanna S. Epp is the Vincent de Paul Professor of Mathematical Sciences at DePaul University. She is the author of *Discrete Mathematics with Applications* and *Discrete Mathematics: An Introduction to Mathematical Reasoning*, as well as many articles about cognitive issues associated with teaching analytical thinking and proof. In 2005, she received the Louise Hay Award for Contributions to Mathematics Education. She enjoys theater, music, opera, movies, and travel, especially trips that involve walking in the mountains.

Erica Flapan is the Lingurn H. Burkhead Professor of Mathematics at Pomona College. Her research is in topology and its applications to chemistry and molecular biology. She has written a book entitled *When Topology Meets Chemistry*, published jointly by Cambridge University Press and the Mathematical Association of America. She has also co-authored a book (together with James Pommersheim and Tim Marks) entitled *Number Theory: A Lively Introduction with Proofs, Applications, and Stories*, published by John Wiley and Sons. In 2011, she won the Mathematical Association of America's Deborah and Franklin Tepper Haimo Award for Distinguished College or University Teaching of Mathematics.

Sol Garfunkel is the executive director of the Consortium for Mathematics and Its Applications and has dedicated the past 35 years to research and development efforts in mathematics education. Garfunkel has been on the mathematics faculty of Cornell University and the University of Connecticut at Storrs. He has served as project director for several National Science Foundation curriculum projects. He was the project director and host for the television series *For All Practical Purposes: Introduction to Contemporary Mathematics*. He is a member of the Mathematics Expert Group of the OECD Programme for International Student Assessment (PISA), and he was the Glenn Gilbert National Leadership Award recipient for 2009 from the National Council of Supervisors of Mathematics.

Andrea Genovese is a lecturer in logistics and supply chain management at the Management School of the University of Sheffield. He holds a Ph.D. in operational research from the University of Naples Federico II (Italy). His education also includes an M.Sc. (with a B.Sc.) in engineering and technology management from the same institution and a master in business administration (MBA) at the Whittemore School of Business and Economics of the University of New Hampshire. He is also acting as a key member for the Logistics and Supply Chain Management (LSCM) Research Centre and for the recently established Centre for Environment Energy and Sustainability (CEES). In recent years, he has contributed to several funded research projects on the topic of low carbon innovation for supply chains. He has published several papers in international journals and regularly presents his work in international conferences.

Bonnie Gold has taught mathematics at Monmouth University in New Jersey since 1998 and before that taught for 20 years at Wabash College in Indiana. She was one of the founders of POM SIGMAA, the special interest group of the Mathematical Association of America (MAA) for the philosophy of mathematics. In 2006, she received the Distinguished Teaching Award from the New Jersey section of the MAA, and in 2012, the Louise Hay Award for Contributions to Mathematics Education from the Association of Women in Mathematics. With Roger Simons, she edited *Proof and Other Dilemmas: Mathematics and*

Philosophy, published by the MAA in 2008. Her "What Is the Philosophy of Mathematics, and What Should It Be?" was published in the *Mathematical Intelligencer* in 1994. She has been involved with the MAA for many years (and is currently a governor representing the New Jersey section) and started section versions of Project NExT in both Indiana and New Jersey.

Fernando Gouvêa is Carter Professor of Mathematics at Colby College. His main academic interests are number theory (especially Galois representations and their connection to modular forms) and the history of mathematics. His writing has won the Bowdoin Prize "for the advancement of useful and polite literature" from Harvard and the Lester R. Ford Award from the Mathematical Association of America. Gouvêa is the author of several books, including *p-adic Numbers: An Introduction* and, jointly with William P. Berlinghoff, *Math through the Ages: A Gentle History for Teachers and Others*, which won the MAA's Beckenbach Book Prize. He was the editor of *MAA Focus* for just more than a decade and is currently the editor of *MAA Reviews*, an online book review service available at http://www.maa.org/maareviews

Timothy Gowers is a Royal Society Research Professor at the University of Cambridge. He is the author of *Mathematics, A Very Short Introduction,* the main editor of *The Princeton Companion to Mathematics,* and the author of several research papers in functional analysis and combinatorics. In 1996, he was awarded a European Mathematical Society Prize, and in 1998 he received a Fields Medal. He is married and has five children.

Jeremy Gray is a professor of the history of mathematics at the Open University, and an honorary professor at the University of Warwick, where he lectures on the history of mathematics. In 2009, he was awarded the Albert Leon Whiteman Memorial Prize of the American Mathematical Society for his work on the history of mathematics. His book *Plato's Ghost: The Modernist Transformation of Mathematics* was published by Princeton University Press in 2008, and his scientific biography of Henri Poincaré will be published by them in November 2012.

Ian Hacking has retired from his chair at the Collège de France, Paris, and from a university professorship at the University of Toronto. He has written many books on many topics, of which his favorite is still *The Emergence of Probability* (1975, 2006). A companion work is *The Taming of Chance* (1990). He has published little on the philosophy of mathematics, but it was his first love. In 2010, he gave the Descartes lectures in Holland and the Howison lectures in Berkeley on this topic. A much extended version of these talks is in press from Cambridge University Press: *The Mathematical Animal: Philosophical Reflections on Proof, Necessity and Human Nature.* In 2009, he was awarded the Holberg

International Memorial Prize given in Norway for "outstanding scholarly work in the fields of the arts and humanities, social sciences, law and theology."

Brian Hayes is a senior writer for *American Scientist* magazine and a former editor of both *American Scientist* and *Scientific American*. He is also the author of *Infrastructure: A Field Guide to the Industrial Landscape* (W. W. Norton, 2005) and *Group Theory in the Bedroom and Other Mathematical Diversions* (Hill and Wang, 2008). He has been journalist-in-residence at the Mathematical Sciences Research Institute in Berkeley and a visiting scientist at the Abdus Salam International Centre for Theoretical Physics in Trieste. He is the winner of a National Magazine Award.

John Huerta received his Ph.D. in 2011 from the University of California. He is currently a researcher at the Australian National University and will be heading in January to Instituto Superior Técnico in Lisbon. He principally studies the mathematics underlying string theory but also has interests across the spectrum of mathematical physics.

Gennaro Improta is a full professor of operational research at the University of Naples Federico II. His main expertise lies in the field of transportation systems design. He has published his work in prestigious international journals like *European Journal of Operational Research, Transportation Research, Operations Research Letters,* and *Computers & Operations Research.*

After 20 years of doing research and development in lasers and optoelectronics, **Robert J. Lang**'s first love, origami, took over his life. Since 2001, he has been a full-time origami artist, consultant, author, and lecturer on the art, its connections to mathematics, science, and technology, and on techniques for origami design. He has more than 80 technical publications and 50 patents in semiconductor lasers and optoelectronics and 14 books and numerous articles in the field of origami. His origami artwork has been exhibited in Tokyo, Paris, Los Angeles, and the Museum of Modern Art in New York and in 2009 was awarded Caltech's highest honor, the Distinguished Alumni Award.

Mario Livio is a senior astrophysicist at the Space Telescope Science Institute, which conducts the scientific program of the Hubble Space Telescope, and will conduct the program for the upcoming James Webb Space Telescope. He is also a best-selling author and a popular lecturer. He has done important research on the topics of accretion onto compact objects, Type Ia Supernovae and their use in cosmology, evolution of binary stars, and extrasolar planets. He is a fellow of the American Association for the Advancement of Science and has received numerous honors, including Carnegie Centenary Professor (2003), AURA Science Award (2005), Danz Distinguished Lecturer (2006), and Resnick Distinguished Lecturer (2006). For his popular book *The Golden Ratio,* Livio received the Peano Prize (2003) and the International Pythagoras Prize (2004). His book *Is God A Mathematician?* was selected by the *Washington Post* as one of the best books of 2009. Livio has a regular blog, *A Curious Mind,* about science, art, and the links between them. His new book, *Brilliant Blunders,* is scheduled to appear in May 2013.

David Mumford is a professor emeritus of applied mathematics at Brown University. He received the Fields Medal for his work in algebraic geometry in 1974 while at Harvard. His applied work is in the mathematics of perception and appears in his book *Pattern Theory,* written jointly with Agnes Desolneux.

Peter Rowlett works for the Maths, Stats and OR Network at the University of Birmingham in the United Kingdom. He is interested in mathematics education and communicating mathematics, and his interest in the history of mathematics arises from its suitability for these activities. He is a member of the Council of the British Society for the History of Mathematics, on whose behalf the "Unplanned Impact of Mathematics" pieces were collected, and an editor of the online mathematics magazine *The Aperiodical*.

Karl Schaffer codirects the Dr. Schaffer and Mr. Stern Dance Ensemble and teaches mathematics at De Anza College, in Cupertino, California. He and co-director Erik Stern have performed their duet concert *Two Guys Dancing About Math* more than 500 times in theaters, schools, and conferences throughout North America since it premiered in 1990. Both he and Stern are on the teaching artist roster of the Kennedy Center's Partners in the Arts program and teach workshops on how to integrate math and dance in the classroom around the country. Their book *Math Dance with Dr. Schaffer and Mr. Stern,* co-authored with Scott Kim, helps teachers find ways to link these subjects. Schaffer's recent dance work includes *The Daughters of Hypatia: Circles of Mathematical Women,* premiered in 2012, which describes the lives, work, and struggles of women mathematicians throughout the ages; and choreography for Keith Devlin's 2010 show *Harmonious Equations,* with world music group Zambra. He has created three outreach math and dance concerts for school-age audiences that are currently performed by dance companies in Colorado Springs and Birmingham, Alabama.

Rob Schneiderman is a professor of mathematics at Lehman College, City University of New York. His sphere of research is centered on the geometric topology of 3- and 4-dimensional manifolds. Before becoming a mathematician, he had a busy career as a jazz musician.

Charlotte Simmons is the associate dean of the College of Mathematics and Science at the University of Central Oklahoma. Trained as a pure mathematician with a Ph.D. in mathematics from the University of Oklahoma, she has had a passion for the history of mathematics and science since her undergraduate days. Her recent publications in this area include articles in the *Bulletin for the British Society for the History of Mathematics,* the *College Mathematics Journal,* and the *Mathematical Intelligencer.* Though she is currently in administration, she continues to involve undergraduates in her research and was the 2012 recipient of the University of Central Oklahoma chapter Sigma Xi Researcher of the Year Award.

David Swart received a master's degree in mathematics from the University of Waterloo and works at Christie Digital in Kitchener, Ontario, where he writes software that warps and blends multiple projectors together into seamless displays. His interests include warping photographic imagery into novel and pleasing art, and he has presented two papers ("Using Turtles and Skeletons to Display the Viewable Sphere" and "Warping Pictures Nicely") at the Bridges Conference, an annual event about art and mathematics.

Terence Tao was born in Adelaide, Australia, in 1975. He completed his Ph.D. under Elias Stein at Princeton in 1996 and has been a professor of mathematics at UCLA since 1999. Tao's areas of research include harmonic analysis, partial differential equations, combinatorics, and number theory. He has received a number of awards, including the Salem Prize in 2000, the Bochner Prize in 2002, the Fields Medal and the SASTRA Ramanujan Prize in 2006, the MacArthur Fellowship and Ostrowski Prize in 2007, the Waterman Award in 2008, the Nemmers Prize in 2010, and the Crafoord Prize in 2012. Tao also currently holds the James and Carol Collins Chair in Mathematics at UCLA and is a fellow of the Royal Society, a corresponding member of the Australian Academy of Sciences, a foreign member of the National Academy of Sciences, and a member of the American Academy of Arts and Sciences.

Bruce Torrence is the Garnett Professor of Mathematics and chair of the Mathematics Department at Randolph-Macon College, where he has been teaching for almost 20 years. In 2008, he received the John Smith Teaching Award from the Mathematical Association of America. He is currently a co-editor of *Math Horizons*.

Notable Texts

The selection for this book is inevitably parsimonious. Perhaps a volume just as interesting and as diverse as this one can be put together from texts I did not include. To guide the reader and to offer further suggestions, I list here articles that, at one time or another, I considered for selecting.

Anthes, Gary. "Nonlinear Systems Made Easy." *Communications of the ACM* 54.1(2011): 17–19.

Alexanderson, Gerald L. "Isaac Newton, Fatio de Duillier, and Alchemy." *Bulletin of the AMS* 48.2(2011): 275–79.

Bombieri, Enrico. "The Mathematical Infinity." In *Infinity: New Research Frontiers*, edited by Michael Heller and W. Hugh Woodin, pp. 55–75. New York: Cambridge University Press, 2011.

Bressoud, David. "Historical Reflections on Teaching the Fundamental Theorem of Integral Calculus." *The American Mathematical Monthly* 118.2(2011): 99–115.

Bueno, Otàvio, and Mark Colyvan "An Inferential Conception of the Applications of Mathematics." *Noûs* 45.2(2011): 345–74.

Butterworth, Brian. "Foundational Numerical Capacities and the Origins of Dyscalculia." *Trends in Cognitive Sciences* 14.11(2010): 534–41.

Darwiche, Adnan. "Bayesian Networks." *Communications of the ACM* 53.12(2011): 80–90.

De Corte, Erik, Lucia Mason, Fien Depaepe, and Lieven Verschaffel. "Self-Regulation of Mathematical Knowledge and Skills." In *Handbook of Self-Regulation of Learning and Performance*, edited by Barry J. Zimmerman and Dale H. Schunk, pp. 155–72. New York: Routledge, 2011.

Edwards, Ann R., Indigo Esmonde, and Joseph F Wagner. "Learning Mathematics." In *Handbook of Research on Learning and Instruction*, edited by Richard E. Mayer and Patricia E. Alexander, pp. 55–77. New York: Routledge, 2011.

Edwards, Michael. "Algorithmic Composition: Computational Thinking in Music." *Communications of the ACM* 54.7(2011): 58–67.

Ehrhardt, Caroline. "Évariste Galois and the Social Time of Mathematics." *Revue d'Histoire des Mathématiques* 17.2(2011): 175–210.

Garfunkel, Sol, and David Mumford. "How to Fix Our Math Education." *The New York Times* August 24, 2011.

Goethe, Norma, and Michèle Friend. "Confronting Ideals of Proof with the Ways of Proving of the Research Mathematician." *Studia Logica* 96.2(2010): 273–88.

Goodwin, Geoffrey P., and P. N. Johnson-Laird. "Mental Models of Boolean Concepts." *Cognitive Psychology* 63(2011): 34–59.

Grattan-Guinness, I. "Numbers as Moments of Multisets: A New–Old Formulation of Arithmetic." *The Mathematical Intelligencer* 33.1(2011): 19–29.

Hayes, Brian. "Quasirandom Ramblings." *American Scientist* 99.4(2011): 282–87.

Hersh, Reuben. "Mathematical Intuition (Poincaré, Polya, Dewey)." *The Montana Mathematics Enthusiast* 8.1-2(2011): 35–50.

Hintikka, Jaakko. "What Is the Axiomatic Method?" *Synthese* 183(2011): 69–85.

Kass, Robert E. "Statistical Inference: The Big Picture." *Statistical Science* 2.16(2011): 1–20.

Katz, Karin U., and Mikhail G. Katz. "Cauchy's Continuum." *Perspectives on Science* 19.4(2011): 426–52.

Kessel, Cathy. "Rumors of Our Rarity are Greatly Exaggerated: Bad Statistics about Women in Science." *Journal of Humanistic Mathematics* 1.2(2011): 2-26.

Klarreich, Erica. "Approximately Hard: The Unique Games Conjecture." (https://simons foundation.org/mathematics-physical-sciences/featured-articles/-/asset_publisher/bo1E/content/approximately-hard-the-unique-games-conjecture?redirect=%2Fmathematics-physical-sciences%2Ffeatured-articles)

Knuth, Donald E. "Mathematical Vanity Plates." *The Mathematical Intelligencer* 33.1(2011): 33–45.

Kvasz, Ladislav. "Kant's Philosophy of Geometry—On the Road to a Final Assessment." *Philosophia Mathematica* 19.2(2011): 139–66.

Leonard, Molly, and Cheri Shakiban. "The Incan Abacus: A Curious Counting Device." *Journal of Mathematics and Culture* 5.2(2010): 81–106.

Lisi, Garrett A., and James Owen Weatherall. "A Geometric Theory of Everything." *Scientific American* 303.6(2010): 55–61.

Maddy, Penelope. "Set Theory as a Foundation." In *Foundational Theories of Classical and Constructive Mathematics*, edited by Giovanni Sommaruga, pp. 85–95. New York: Springer, 2011.

Marfori, Marianna Antonutti. "Informal Proof and Mathematical Rigour." *Studia Logica* 96.2(2010): 261–72.

Mauldin, Tim. "Time, Topology, and Physical Geometry." *Proceedings of the Aristotelian Society (Supplementary Volume)* 84(2010): 63–78.

Mazliak, Laurent, and Glenn Shafer. "What Does the Arrest and Release of Emile Borel and His Colleagues in 1941 Tell Us about the German Occupation of France?" *Science in Context* 24.4(2011): 587–623.

Roberts, Bryan W. "How Galileo Dropped the Ball and Fermat Picked It Up." *Synthese* 180(2011): 337–56.

Salerno, Adriana. "Partition Numbers Unveiled as Fractal." *MAA Focus* 31.2(2011): 5, 7.

Spaepen, Elizabet, Marie Coppola, Elizabeth S. Spelke, Susan E. Carey, Susan Goldin-Meadow. "Number Without a Language Model." *Proceedings of the National Academy of Sciences* 108.8(2011): 3163–68.

Sykes, Geoffrey. "Media as Mathematics: Calculating Justice." In *Courting the Media: Contemporary Perspectives on Media and Law*, edited by Geoffrey Sykes, pp. 127–47. New York: Nova Science Publishers, 2010.

Tall, David. "Crystalline Concepts in Long-Term Mathematical Invention and Discovery." *For the Learning of Mathematics* 31.1(2011): 3–8.

Twigg , Carol A. "The Math Emporium: A Silver Bullet for Higher Education." *Change: The Magazine for Higher Learning* 43.3(2011): 25–34.

Wangler, Tom. "An Irrational Conundrum." *Math Horizons* 19.1(2011): 9–11.

Webb, Richard. Spaghetti Functions. *New Scientist* October 15, 2011.

Weber, Matthias, and Michael Wolf. "Early Images of Minimal Surfaces." *Bulletin of the AMS* 48.3(2011): 457–60.

Weinberg, Steven. "Symmetry: A 'Key to Nature's Secrets.'" *New York Review of Books* October 27, 2011.

Wu, H. "The Mis-Education of Mathematics Teachers." *Notices of the American Mathematical Society* 58.3(2011): 372–84.

Yau, Shing-Tung, and Steve Nadis. "String Theory and the Geometry of the Universe's Hidden Dimensions." *Notices of the American Mathematical Society* 58.8(2011): 1067–76.

Young, Stanley S., and Alan Karr. "Deming, Data, and Observational Studies." *Significance* 6(2011): 116–20.

Acknowledgments

I am grateful to the authors and to the original publishers of the texts reprinted in this anthology for cooperating in the tasks needed for the reissue in this form.

Gerald Alexanderson and Fernando Gouvêa indirectly helped me reach the final selection by sending detailed comments on a broader collection of texts; yet the responsibility for the eventual shortcomings in the final result is solely mine. Fernando Gouvêa also suggested a long list of potential candidates; several of the articles he advanced were under consideration and are included in this book. Thanks to Paul Campbell for useful observations on the previous volumes and to Allen Knutson for a marketing tip. Audiences at Cornell University in Ithaca and California Polytechnic in San Luis Obispo helped me understand better the readers' perspective on this enterprise.

As usual, I received support and advice from the in-house editor at Princeton University Press, Vickie Kearn. Quinn Fusting took care of the copyright issues, Nathan Carr oversaw the production process, and Paula Bérard copyedited the manuscript. Thank you to all.

Editing this annual series while teaching intensively and researching and writing for a doctoral degree is no easy feat. I could not have accomplished it without my advisors' understanding; thanks to my committee members at Cornell University, professors David W. Henderson (chair), Anil Nerode, and John Sipple. Special thanks to Maria Terrell for assigning me an adequate teaching load—and to Dan Barbasch, Mary Ann Huntley, Severin Drix, Michelle Klinger, Catherine Penner, and Abby Eller for co-opting me in supplementary teaching with the Cornell Mathematics Outreach Programs, Cornell Summer School, and Cornell Adult University. Also thanks to Katherine Gottschalk, Paul Sawyer, and David Faulkner of the Cornell Knight Institute for Writing in the Disciplines, for giving me the chance to teach again my Writing in Mathematics seminar.

Thanks to the Cornell University Library; its staff, services, acquisitions, collections, and subscriptions are indispensable to my work on this series.

Fangfang Li was unequaled in patience, kindness, and companionship.

In a good part of my apartment, my daughter Ioana has to take care not to mix up, spread out, or step on countless piles of books and articles. Thanks for making sure that I can work even after you run, jump, and tumble around our crowded place.

I dedicate this volume to my parents. They spared no effort in sending me from a small village "to the city," so that I could attend better schools—thus setting me on a path of opportunities richer than they dreamed.

Credits